Advances in Bioceramics
and Biocomposites

Advances in Bioceramics and Biocomposites

*A collection of papers presented at the
29th International Conference
on Advanced Ceramics and Composites,
January 23-28, 2005,
Cocoa Beach, Florida*

Editor
Mineo Mizuno

General Editors
Dongming Zhu
Waltraud M. Kriven

Published by

The American Ceramic Society
735 Ceramic Place
Suite 100
Westerville, Ohio 43081
www.ceramics.org

Advances in Bioceramics and Biocomposites

For information on ordering titles published by The American Ceramic Society, or to request a publications catalog, please call 614-794-5890, or visit www.ceramics.org

ISSN 0196-6219

ISBN 1-57498-236-2

Contents

Preface

A *"Bioceramics and Biocomposites"* session was started in 2002 in the 26th International Conference on Advanced Ceramics and Composites. The session was raised to a Bioceramic symposium in 2005. We appreciate the program chair for the decision. It was timely since bioceramics have been recognized to be one of the most important materials in order to overcome problems of an aging society in the near future.

The use of ceramics in biological environments and biomedical applications is of increasing importance, as is the understanding of how biology works with minerals to develop strong materials. Bones and teeth are composed of inorganic (calcium phosphate) and organic (protein) materials. They are ultimate composites, being skillfully tailored to show both structural and bioactive functions. Therefore this symposium contained several topics, such as biomimetics, processing of materials for biomedical applications, interactions of ceramics in biological/biomedical applications, performance issues in biomedical ceramics, orthopaedic replacements, and dental ceramics. A total of 45 papers were presented in this Symposium: including 10 invited papers. Authors, from academia, national laboratories, industries, and government agencies, gathered in Cocoa Beach in Florida in 2005, from 9 countries around the world.

The symposium organizers would like to thank all of the participants in the symposium and the staff at the ACerS. We appreciate ACerS for their efforts in organizing the review process and coordinating the production of this volume of Ceramic Engineering and Science Proceedings. The symposium organizers hope that this symposium will promote the quality of life for humanity.

Mineo Mizuno
Jian Ku Shang
Richard Rusin
Waltraud Kriven
Besim Ben-Nissan

Processing of Biomaterials

PREPARATION AND BIOACTIVE CHARACTERISTICS OF POROUS BORATE GLASS SUBSTRATES

Mohamed N. Rahaman, Wen Liang, and Delbert E. Day, University of Missouri-Rolla, Department of Materials Science and Engineering, and Materials Research Center, Rolla, MO 65409

Nicholas W. Marion, Gwendolen C. Reilly, and Jeremy J. Mao, University of Illinois at Chicago, Department of Bioengineering and Tissue Engineering Laboratory, Chicago, IL 60607

ABSTRACT

Whereas silicate-based bioactive glasses and glass-ceramics have been widely investigated for bone repair or as scaffolds for cell-based bone tissue engineering, recent data have demonstrated that silica-free borate glasses also exhibit bioactive behavior. The objectives of this study were to fabricate porous, three-dimensional substrates of a borate glass and to investigate the biocompatibility of the borate glass substrates by *in vitro* cell culture with human mesenchymal stem cells (hMSCs) and hMSC-derived osteoblasts (hMSC-Obs). Borate glass particles with sizes 212-355 μm were loosely compacted and then sintered at 600°C to form porous disc-shaped substrates (porosity ≈ 40%). Partial or nearly complete conversion of the glass substrates to a calcium phosphate (Ca-P) material was achieved by soaking the substrates for 1 day or 7 days in a 0.25 molar K_2HPO_4 solution at 37°C and at pH of 9.0. Bone marrow derived hMSCs and hMSC-Obs seeded in the samples both adhered to the porous constructs whereas hMSC-Obs markedly synthesized alkaline phosphatase, an early osteogenic marker. These data indicate strong bioactive characteristics for the borate glass constructs and the potential use of the constructs for bone tissue engineering.

INTRODUCTION

Certain compositions of glasses, glass-ceramics, and ceramics, referred to as bioactive ceramics, have been widely investigated for healing bone defects, due to their ability to enhance bone formation and to bond to surrounding tissue [1-5]. Cell-seeded bioactive ceramics are also of interest as potential scaffolds for bone tissue engineering [6,7]. Hydroxyapatite and tricalcium phosphate ceramics, composed of the same ions as bone, are biocompatible and produce no systemic toxicity or immunological reactions, but they resorb slowly or undergo little conversion to a bone-like material after implantation [8,9]. Many bone regeneration applications require gradual resorption of the implanted biomaterials and concurrent replacement of the biomaterials by the host bone.

Bioactive glasses are superior to the less reactive ceramics in that they are osteoinductive as opposed to osteoconductive. Furthermore, the dissolution and conversion of bioactive glasses to a calcium phosphate (Ca-P) material seems to induce bone cell differentiation [10]. A characteristic feature of bioactive glasses is the time-dependent modification of the surface, resulting in the formation of a calcium phosphate (Ca-P) layer through which a bond with the surrounding tissue is established [11,12]. It has been suggested that the formation of a Ca-P layer *in vitro* is indicative of a material's bioactive potential *in vivo* [4,5,13].

Since the report of its bone bonding properties in 1971 by Hench *et al.* [14], the bioactive glass codenamed 45S5, referred to as Bioglass®, with the composition of 45% SiO_2, 6% P_2O_5, 24.5% Na_2O, and 24.5% CaO (by weight), has received most interest for biological applications [4,5]. Bioactive glasses based on the 45S5 composition are attractive scaffold materials because their rapid bonding to bone provides early mechanical stability, in addition to stimulating osteo-progenitor cell function, and biocompatibility [15-17]. *In vivo* studies have shown that 45S5 glass can stimulate bone regeneration [18-20], whereas *in vitro* studies have shown that the glass itself and the soluble ionic species released by dissolution have an osteoinductive effect [21-24]. Porous bioactive silicate glass constructs based on the 45S5 composition have been developed as possible tissue engineering scaffolds [25,26]. Cell culture experiments indicated that the porous glass can function as a template for generating mineralization *in vitro* [25].

The low chemical durability of some borate glasses has been known for decades but the potential of borate glasses in biomedical applications has not been explored until recently [27,28]. A borate glass, designated 45S5B1, with the same composition as 45S5 bioactive glass but with all the SiO_2 replaced by B_2O_3, was investigated by Richard [29]. *In vitro* experiments indicated that a Ca-P layer forms on the surface of the borate glass upon immersion in a K_2HPO_4 solution at 37°C and that the Ca-P layer forms more rapidly on the borate glass than on 45S5 bioactive glass [29]. As a first *in vivo* experiment, 45S5B1 borate glass particles (partially reacted in a K_2HPO_4 solution to produce a surface Ca-P layer) and 45S5 glass particles were separately implanted into defects (0.6–1.2 mm in diameter) in the tibia of rats [29]. Histological examination of the harvested constructs indicated that the partially converted borate glass particles promoted bone growth more rapidly than the 45S5 glass particles. Both types of glass particles promoted sufficient bone growth for closure of the implant site after 60 days [29].

The more rapid conversion of borate glass to Ca-P at near body temperature and the favorable *in vivo* reaction of particles to produce bonding with bone warrant additional investigations of the value of borate glass as bone replacement materials and as scaffolds for bone tissue engineering. However, little is known about the fabrication of the borate glass into porous, three-dimensional constructs or the effects of the borate glass on cell attachment, growth and differentiation. The objectives of this study were to produce porous, three-dimensional substrates of a borate glass intended for bone tissue engineering and to investigate the effects of the fabricated borate glass constructs on attachment and differentiation of human mesenchymal stem cells (hMSCs) and hMSC-derived osteoblasts (hMSC-Obs).

EXPERIMENTAL PROCEDURE
Fabrication of Borate Glass Substrates

Particles of borate glass (Na_2O-CaO-B_2O_3) were prepared by melting reagent grade chemicals in a platinum crucible, quenching the melt, and crushing the glass in a hardened steel mortar and pestle. After removing the metallic impurities magnetically, the particles were sieved through stainless steel sieves to produce sizes in the range of 212–355 µm. Porous disc-shaped substrates (15 mm diameter × 2–3 mm thickness) were produced by pouring the glass particles into vibrating graphite molds, followed by sintering for 10 min at 600°C. The structure of the porous substrates was examined using X-ray diffraction and optical microscopy. The porosity of the substrates was estimated from the computer imaging of optical micrographs and from the measured density.

Conversion of the porous borate glass substrates to Ca-P was investigated by immersing the substrates in 0.25 molar K_2HPO_4 solution with a starting pH value of 9.0 at 37°C and measuring the weight loss as a function of time. The structural characteristics of the converted material were observed using scanning electron microscopy (SEM; Hitachi S-4700). Some glass substrates used in cell culture experiments were partially or fully converted to Ca-P to determine the most favorable condition of the borate glass for supporting cell growth and differentiation. The partially converted borate glass substrates (denoted pBG) and the fully converted substrates (denoted Ca-P) were prepared by immersing the porous glass substrates for 1 day and 7 days, respectively, in the K_2HPO_4 solution.

Cell Culture on Porous Borate Glass Substrates
    Human bone marrow derived mesenchymal stem cells (hMSCs) were isolated from bone marrow samples (AllCells, Berkeley, CA) using a RosetteSep kit (Stem Cell Technologies, Inc., Vancouver, BC, Canada). The hMSCs were grown in monolayer in cell culture media consisting of 89% DMEM, 10% FBS, 1% penicillin – streptomycin (basal cell culture media). After 4 days non-adherent cells were removed and the media was changed every 4 days. Cells were passaged up to four times each time upon confluency. Upon the 4th passage, 50% of the hMSCs were exposed to osteogenic supplemented medium (basal cell culture media, 100 nM dexamethasone, 50 μg/mL L-ascorbic acid-2-phosphate). Upon exposure to osteogenic supplement, hMSCs differentiated into osteoblastic cells (hMSC-Obs) [30-32], whereas the other 50% hMSCs continued incubation in basal culture complete media without osteogenic supplement.
    The hMSCs and hMSC-Obs were seeded (30,000 cells per cm^3) on porous substrates of the unconverted borate glass (BG), the partially converted borate glass (pBG), or the completely converted borate glass (Ca-P), and incubated for an additional 14 days. Live cell assay was then performed using Promega (Madison, WI) CellTiter 96® AQ$_{ueous}$ One Solution Cell Proliferation Assay, which quantified cell viability through NADH activity using 3-(4,5-dimethyl-2-yl)-5-(3-carboxymethoxyphenyl)-2-(4-sulfophenyl)-2H-tetrazolium, inner salt (MTS). The absorbance values for MTS correlate with a live cell number as documented in the product information sheet. Alkaline phosphatase activity (AP) was assayed by Napthol as-biphosphate, fast red violet salt, and N,N dimethylformamide solution (Sigma-Aldrich Co., St. Louis, MO).

RESULTS AND DISCUSSION
    Figure 1 shows an optical micrograph of the surface of a porous borate glass substrate produced by sintering. The touching particles are bonded at the necks, providing enhanced strength without significant flow of the glass into the pores. The reduction of the porosity of the substrates during sintering was negligible. Computer imaging of optical micrographs indicated that the substrates had a porosity of 40-45% and a median pore size of 100-150 μm. The porosity estimated by computer imaging was in agreement with the value determined from the measured density of the substrate and the density of the fully dense glass (2.58 g/cm^3). X-ray diffraction showed that the glass in the porous substrate remained amorphous after sintering.

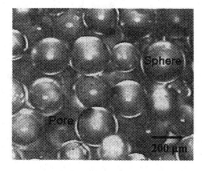

Figure 1. Optical micrograph of the surface of a porous borate glass substrate produced by sintering a loosely compacted mass of particles (212-355 μm) for 10 min at 600°C.

The weight loss data for the porous borate glass substrates during their conversion to Ca-P in K_2HPO_4 solution are shown in **Fig. 2** as a function of time. Conversion of the glass to Ca-P, as indicated by the maximum weight loss (60-65%), was completed after approximately 7 days. The conversion of the borate glass to Ca-P is believed to involve dissolution of the glass into the surrounding liquid and precipitation of calcium and phosphate ions onto the surface of the sub-strate [33]. Assuming that all of the sodium and borate ions from the glass go into solution and all of the calcium ions go into the formation of a Ca-P material with the composition of stoichiometric hydroxyapatite, $Ca_{10}(PO_4)_6(OH)_2$, then the theoretical weight loss should be 69%. The discrepancy between the maximum measured weight loss and the theoretical weight loss may be due to incomplete conversion of the glass, some calcium ions remaining in solution, the formation of a nonstoichiometric hydroxyapatite with a Ca/P ratio lower than the stoichiometric value of 1.67, or a combination of all three factors. Chemical analysis of the Ca-P material formed by the conversion of similar borate glasses under the same conditions indicated that the Ca/P ratio was well below 1.67 [34].

Conversion of the borate glass to Ca-P starts at the surface and moves inward [33]. By con-trolling the time of reaction in the K_2HPO_4 solution, substrates with different ratios of the sur-rounding Ca-P layer to the borate glass core can be produced. Constructs reacted for 1 day con-sisted of an interconnected mass of composite particles, with a thin surface layer of the glass converted to Ca-P. The thickness of the Ca-P layer, estimated from the weight loss data was 40-50 μm. Substrates reacted in the K_2HPO_4 solution for 7 days were almost fully converted to Ca-P, and consisted of an interconnected mass of Ca-P particles. **Figure 3** shows SEM micrographs of the surfaces of the three types of porous substrates used in the present work in cell culture ex-periments. The unconverted borate glass (BG) substrate has smooth surfaces characteristic of the spheroidized glass particles, whereas the constructs of the partially converted glass (pBG) and the fully converted glass (Ca-P) have less smooth surfaces. High resolution SEM, performed in related work [35], indicated that the Ca-P material was highly porous, with fine pores on the or-der of several tens of nanometers.

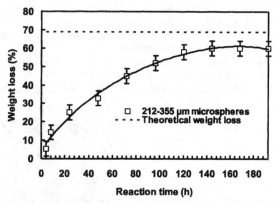

Figure 2. Weight loss of porous borate glass substrates as a function of time in 0.25 molar K_2HPO_4 solution at 37°C and a pH value of 9.0. Conversion of the glass to a calcium phosphate (Ca-P) material in the solution is accompanied by a weight loss. The estimated theoretical weight loss is shown by the horizontal dotted line.

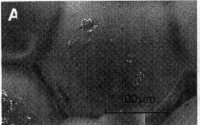

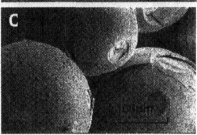

Figure 3. SEM micrographs of the of the surfaces of borate glass substrates used in cell culture experiments: (A) unconverted borate glass (BG); (B) partially converted borate glass (pBG) formed by reaction for 1 day in K_2HPO_4 solution; (C) fully converted borate glass (Ca-P) formed by reaction for 7 days in K_2HPO_4 solution.

The differences in the condition of the borate glass substrates may influence the interaction with cells. However, the most favorable condition of the borate glass for cellular interaction is, at present, unclear. The unconverted borate glass (BG) with its smooth surface initially may not provide favorable sites for cell attachment and significant dissolution of calcium, sodium and borate ions will occur initially into the surrounding fluid as the glass surface reacts with the fluid. For constructs of the partially converted glass (pBG), the porous Ca-P surfaces may provide more favorable sites for cell attachment. Dissolution of calcium, sodium and borate ions into the surrounding fluid is still expected to occur but at a lower rate than for the unconverted glass. The fully converted constructs (Ca-P) provide surface sites similar to those of the pBG constructs, but almost no dissolution of sodium and borate ions into the surrounding fluid will occur due to the absence of any significant quantity of borate glass in the substrate.

The unconverted borate glass substrates (BG) disintegrated during cell culture experiments, presumably due to reactions of the glass with the cell culture medium. However, the partially converted substrates (pBG) and the fully converted substrates (Ca-P) remained intact and maintained their original cylindrical shape throughout the experiments. Live cell number (MTS) assayed after 14 days verified the cell viability of both hMSCs and hMSC-Obs cultured on the pBG and Ca-P substrates. The hMSCs seeded on the pBG templates had significantly higher cell viability than hMSCs seeded on the Ca-P templates (**Fig. 4**). The data show a similar trend for hMSC-Obs seeded on the pBG and Ca-P templates but the difference is not significant due to the wider variability of the data for Ca-P templates. The higher cell viability of the hMSC on the pBG substrates may indicate that pBG stimulates cell function. As outlined earlier, a key difference between the pBG and Ca-P substrates is the potential for dissolution of calcium, sodium, and borate ions from the underlying borate glass core of the pBG templates into the culture medium. The mechanism by which these ions may influence cell function is not clear at present but may be important for determining the optimum condition of the borate glass substrates for tissue engineering applications. For cells seeded on the pBG substrates, the data in **Fig. 4** also indicated that the osteogenic cells had significantly higher cell viability than hMSCs. The hMSCs were initially seeded at a higher density than the hMSC-Obs (9,000 cells per construct, versus 3,000 cells per construct, respectively) due to the fact that MSCs proliferated more rapidly than MSC-Obs during the pre-treatment. The data presented in **Fig. 4** plots the MTS absorbance per seeded cell number normalized to MSCs grown on Ca-P. Future studies will investigate apoptosis and proliferation of MSCs and their differentiated osteoblasts on the borate glass substrates.

Active alkaline phosphatase was produced by the cells within the borate glass substrates as indicated by the dark red stain (**Fig. 5**). Higher alkaline phosphatase activity was seen in hMSC-Obs samples (**Fig. 5C, D**) as compared to the undifferentiated, hMSCs (**Fig. 5A, B**). This indicates that a combination of osteogenic supplements with a bioactive borate glass substrate will have a positive effect on osteogenic differentiation.

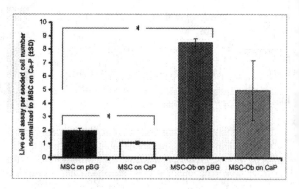

Figure 4. Live cell assay using light absorbance as an metabolic indicator of hMSCs and hMSC-Obs on partially converted borate glass (pBG) and Ca-P substrates. n=4; *= Students' T-test p < 0.05.

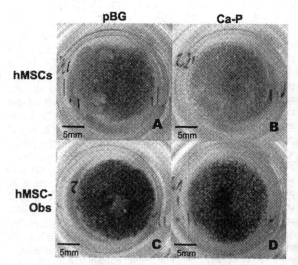

Figure 5. Photographs of (A) human mesenchymal stem cells (hMSCs) cultured on substrates of partially converted borate glass (pBG), (B) hMSCs cultured on substrates of fully converted borate glass (Ca-P), (C) hMSC derived osteoblasts (hMSC-Obs) cultured on pBG substrates, and (D) hMSC-Obs cultured on Ca-P substrates. All samples stained for alkaline phosphatase activity using the same protocol. Red (dark) stain indicates high alkaline phosphatase activity.

8

CONCLUSIONS

Porous bioactive borate glass substrates, prepared by sintering a loosely compacted mass of particles, were conditioned in 0.25 molar K_2HPO_4 solution at 37°C to convert a controlled amount of the glass to a calcium phosphate (Ca-P) material. The cytocompatibility of porous substrates consisting of the unconverted glass (BG), the partially converted glass (pBG), and the fully converted glass (Ca-P), was investigated by *in vitro* cell culture with human mesenchymal stem cells (hMSCs) and hMSC derived osteoblasts (hMSC-Obs). The hMSCs seeded on pBG substrates had a higher metabolic activity and cell viability than on Ca-P substrates. For pBG substrates, hMSC-Obs had significantly higher cell viability than hMSCs. Alkaline phosphatase activity on the pBG and Ca-P substrates with hMSC-Obs revealed the ability of these materials to support osteogenic cells. The data suggest the necessity for additional *in vitro* and *in vivo* investigations of the potential of bioactive borate glass as a cell-accommodating scaffold for bone tissue engineering. In particular, the pBG construct, consisting of a network of borate glass particles surrounded by a Ca-P layer, had the highest cell viability for both cell types and may represent a more favorable condition of the borate glass for bone tissue engineering.

ACKNOWLEDGEMENTS

The presented research was supported by a University of Missouri Research Board Grant (to M.N.R.), a Biomedical Engineering Research Grant from the Whitaker Foundation RG-01-0075, IRIB Grant on Biotechnology jointly from the University of Illinois at Chicago) and the University of Illinois at Urbana-Champaign, and by Research Grants DE13964, DE15391, and EB02332 from the National Institutes of Health (to J.J.M.).

REFERENCES

[1] L. L. Hench and J. Wilson, "Surface Active Biomaterials," Science, 226, 630-636 (1984).

[2] T. Yamamuro, L. L. Hench, and J. Wilson, Eds. Handbook of Bioactive Ceramics, Vols. 1: Bioactive Glasses and Glass-Ceramics. Boca Raton, FL: CRC Press (1990).

[3] T. Yamamuro, L. L. Hench, and J. Wilson, Eds. Handbook of Bioactive Ceramics, Vol. 2: Calcium Phosphate and Hydroxylapatite Ceramics. Boca Raton, FL: CRC Press (1990).

[4] L. L. Hench, "Bioceramics: From Concept to Clinic," J. Am. Ceram. Soc., 74, 1487-1510 (1991).

[5] L. L. Hench, "Bioceramics," J. Am. Ceram. Soc., 81, 1705-1728 (1998).

[6] S. A. Goldstein, P. V. Patil, and M. R. Moalli, "Perspectives on Tissue Engineering of Bone," Clin. Orthop., 367S, S419-S423 (1999).

[7] J. M. Karp, P. D. Dalton, and M. S. Shoichet, "Scaffolds for Tissue Engineering," MRS Bulletin, 28, 301-306 (2003).

[8] C. Klein, P. Patka, and W. den Hollander, "Macroporous Calcium Phosphate Bioceramics in Dog Femora: A Histological Study of Interface and Biodegradation," Biomaterials, 10, 59-62 (1989).

[9] R. B. Martin, M. W. Chapman, N. A. Sharkey, S. L. Zissimos, B. Bay, and E. C. Shor, "Bone Ingrowth and Mechanical Properties of Coralline Hydroxyapatite 1 yr after Implantation," Biomaterials, 14, 341-348 (1993).

[10] L. L. Hench, I. D. Xynos, A. J. Edgar, L. D. K. Buttery, and J. M. Polak, "Gene Activating Glasses," In: Proc. Int. Congr. Glass, Vol. 1. Edinburg, Scotland, 1-6 July, 2001; pp. 226-233.

[11] L. L. Hench and H. A. Paschall, "Direct Chemical Bonding between Bioactive Glass-ceramic Materials and Bone," J. Biomed. Mater. Res. Symp., 4, 25-42 (1973).

[12] T. Kokubo, S. Ito, Z. T. Huang, T. Hayashi, S. Sakka, T. Kitsugi, and T. Yamamuro, "Ca-P-rich Layer Formed on High Strength Bioactive Glass-ceramic A-W," J. Biomed. Mater. Res., 24, 331-343 (1990).

[13] P. Ducheyne, "Bioceramics: Material Characteristics versus *in vivo* Behavior," J. Biomed. Mater. Res., 21, 219-236 (1987).

[14]L. L. Hench, R. J. Splinter, W. C. Allen, and T. K. Greenlee, Jr., "Bonding Mechanisms at the Interface of Ceramic Prosthetic Materials," J. Biomed. Mater. Res., 2, 117-141 (1971).

[15]P. Ducheyne, A. El-Ghannam, and I. M. Shapiro, "Effect of Bioactive Glass Templates on Osteoblast Proliferation and *in vitro* Synthesis of Bone-like Tissue," J. Cell. Biochem., 56, 162-167 (1994).

[16]P. Ducheyne, "Stimulation of Biological Function with Bioactive Glass," MRS Bulletin, 23, 43-49 (1998).

[17]A. El-Ghannam, P. Ducheyne, and I. M. Shapiro, "Effect of Serum Protein Adsorption on Osteoblast Adhesion to Bioglass and Hydroxyapatite," J. Orthop. Res., 17, 340-345 (1999).

[18]D. L. Wheeler, K. E. Stokes, H. M. Park, and J. O. Hollinger, "Evaluation of Particulate Bioglass® in a Rabbit Radius Ostectomy Model," J. Biomed. Mater. Res., 35, 249-254 (1997).

[19]D. L. Wheeler, K. E. Stokes, R. G. Hoellrich, D. L. Chamberland, and S. W. McLoughlin, "Effect of Bioactive Glass Particle Size on Osseous Regeneration of Cancellous Defects," J. Biomed. Mater. Res., 41, 527-533 (1998).

[20]H. Oonishi, L. L. Hench, J. Wilson, F. Sugihara, E. Tsuji, S. Kushitani, and H. Iwaki, "Comparative Bone Growth Behavior in Granules of Bioceramic Materials of Various Sizes," J. Biomed. Mater. Res., 44, 31-43 (1999).

[21]E. A. B. Effah Kaufmann, P. Ducheyne, and I. M. Shapiro, "Evaluation of Osteoblast Response to Porous Bioactive Glass (45S5) by RT-PCR Analysis," Tissue Eng., 6, 19-28 (2000).

[22]I. A. Silver, J. Deas, and M. Erecińska, "Interactions of Bioactive Glasses with Osteoblasts *in vitro*: Effects of 45S5 Bioglass®, and 58S and 77S Bioactive Glasses on Metabolism, Intracellular Ion Concentrations and Cell Viability," Biomaterials, 2001; 22:175-185.

[23]I. D. Xynos, M. V. J. Hukkanen, J. J. Batten, L. D. Buttery, L. L. Hench, and J. M. Polak, "Bioglass® 45S5 Stimulates Osteoblast Turnover and Enhances Bone Formation *in vitro*: Implications and Applications for Bone Tissue Engineering," Calcif. Tissue Int., 67, 321-329 (2000).

[24]I. D. Xynos, A. J. Edgar, L. D. K. Buttery, L. L. Hench, and J. M. Polak, "Gene-expression Profiling of Human Osteoblasts Following Treatment with the Ionic Products of Bioglass® 45S5 Dissolution," J. Biomed. Mater. Res., 55, 151-157 (2001).

[25]A. El-Ghannam, P. Ducheyne, and I. M. Shapiro, "A Bioactive Glass Template for the *in vitro* Synthesis of Bone," J. Biomed. Mater. Res., 29, 359-370 (1995).

[26]E. A. B. Effah Kaufmann, P. Ducheyne, and I. M. Shapiro, "Evaluation of Osteoblast Response to Porous Bioactive Glass (45S5) Substrates by RT-PCR Analysis," Tissue Eng., 6, 19-28 (2000).

[27]D. E. Day, J. E. White, R. F. Brown, and K. D. McMenamin, "Transformation of Borate Glasses into Biologically Useful Materials," Glass Technology, 44, 75-81 (2003).

[28]S. D. Conzone, R. F. Brown, D. E. Day, and G. J. Ehrhardt, "*In vitro* and *in vivo* Dissolution Behavior of a Dysprosium Lithium Borate Glass Designed for the Radiation Synovectomy Treatment of Rheumatoid Arthritis," J. Biomed. Mater. Res., 60, 260-268 (2002).

[29]M. N. C. Richard, Bioactive Behavior of a Borate Glass. M.S. Thesis, University of Missouri-Rolla, 2000.

[30]A. I. Caplan, "Mesenchymal Stem Cells," J. Orthop. Res., 9, 641-650 (1991).

[31]M. F. Pittenger, A. M. Mackay, S. C. Beck, R. K. Jaiswal, R. Douglas, J. D. Mosca, M. A. Moorman, D. W. Simonetti, S. Craig, and D. R. Marshak, "Multilineage Potential of Adult Human Mesenchymal Stem Cells," Science, 284, 143-147 (1999).

[32]A. Alhadlaq, and J. J. Mao, "Tissue-engineered Neogenesis of Human-shaped Mandibular Condyle from Rat Mesenchymal Stem Cells," J. Dent. Res., 82, 950-955 (2003).

[33]J. A. Wojcik, Hydroxyapatite Formation on a Silicate and Borate Glass. M.S. Thesis, University of Missouri-Rolla, 1999.

[34]X. Han, Reaction of Sodium Calcium Borate Glass to Form Hydroxyapatite and Preliminary Evaluation of Hydroxyapatite Microspheres used to Absorb and Separate Proteins. M.S. Thesis, University of Missouri-Rolla, 2003.

[35]W. Liang, N. W. Marion, G. C. Reilly, D. E. Day, J. J. Mao, and M. N. Rahaman, "Bioactive Borate Glass as a Scaffold Material for Bone Tissue Engineering," Submitted to J. Biomed. Mater. Res. (2004).

PROCESSING OF THERMALLY SPRAYED TRICALCIUM PHOSPHATE (TCP) COATINGS ON BIORESORBABLE POLYMER IMPLANTS

M. Baccalaro[a], R. Gadow, A. Killinger*, K. v. Niessen
University of Stuttgart
Allmandring 7b
Stuttgart, D-70569, Germany

ABSTRACT

Bioresorbable polymer implants are a promising research and development field in maxillofacial surgery, since their use eliminates the need for a secondary operation to remove metal implants. Mechanical properties and biocompatibility of these implants are nevertheless not completely satisfactory, and new composite devices are demanded. Thermally sprayed tricalcium phosphate (TCP) coatings may significantly increase the biocompatibility of polymer implants and contribute to match the resorption rate of the device with the bone healing rate, leading to a correct mechanical stress transfer implant/tissue and therefore to successful fracture fixation even in load conditions. The manufacturing process of suitable β-TCP powders for the Atmospheric Plasma Spraying (APS) via spray-drying granulation and the thermal spray coating process on bioresorbable poly (D,L) lactide devices are reported and discussed.

INTRODUCTION

In the last decades many efforts were made to develop biodegradable implants for internal fixation of fractured bones in order to substitute the normally employed metallic implants, since these implants must be removed, if possible, after the successful healing and union of the tissues concerned. Today the mostly employed and studied bioresorbable materials are probably tricalcium phosphate [TCP, $Ca_3(PO_4)_2$] and hydroxylapatite [HAP, $Ca_{10}(PO_4)_6(OH)_2$] among ceramic materials [1, 2] and polyglycolic (PGA) or polylactid (PLA) acids among polymers [3, 4]. Calcium phosphates are generally brittle materials, which are employed only in form of powder (TCP) as bone filler material or as coating on metallic implants (HAP). The main advantage of calcium phosphates is that they are the most important inorganic constituents of biological hard tissues, such as bones. In particular up to 70 % (wt.) of bones is composed of natural HAP, and up to 99 % (wt.) in case of teeth [5]. Synthetic HAP is a bioactive osteoconductive material, able to create a strong chemical bond with the surrounding tissue, and when implanted in the human body it remains stable and chemically unchanged over years. On the other hand TCP is a bioresorbable, osteoinductive ceramic, which is gradually resorbed and replaced in an human body by new bone according to (1):

$$4Ca_3(PO_4)_2 + 2H_2O \rightarrow Ca_{10}(PO_4)_6(OH)_2 + 2Ca^{2+} + 2HPO_4^{2-} \quad (1)$$

TCP is known as existing in two polymorphs, the β-phase (rhombohedral) stable up to 1120 °C, the α and α' - phases (monoclinic) stable in the range 1120 °C-1820 °C. The α-phase can be found at room temperature only as a metastable phase. Previous studies pointed out that α-TCP has a much higher resorption rate in the physiological environment than β-TCP [6].

Polymer fracture fixation devices present some advantages compared to metal implants in particular regarding their mechanical properties, because they match the natural properties of bones much better than metal fixation devices. Polymers are less stiff than metals and this

---

[a] now with Robert-Bosch GmbH Stuttgart

prevents the possible atrophy of bones due to stress protection by the rigid metal osteosynthesis plate. Besides bioresorbable polymer implants prevent the secondary operative procedure to remove the metallic implant. Another advantage of bioabsorbable polymer devices is that they can be shaped and bent to meet geometric conditions when heated to a temperature near the glass transition point (T_g) (f. e. done by the surgeon with the help of a heat gun during implant surgery outside the body). An ideal bioresorbable fixation implant should be degraded during the bone healing process at the same rate at which the bone repairing occurs [7, 8]. This would ensure a correct stress transfer, i.e. a gradual load transmission from the resorbing implant, which is slowly loosening its mechanical properties during the resorption, to the healing bones, which are gradually restoring the original mechanical characteristics previous to the fracture. In maxillofacial surgery one of the most recently applied polymer is amorphous poly (D,L) lactide (PDLLA). This material generally doesn't cause any foreign body reaction during resorption and it is employed to produce plates and screws for fracture fixation of mechanically unloaded bony fragments [9]. The application in loaded zones (e.g. in jaw bone fractures) may nevertheless lead to failure of the implant itself. This is due to the resorption rate of the material, which occurs faster than the healing process of the fracture. The mismatch between resorption rate and healing rate leads therefore to an incorrect stress transfer. Another point is that the implant's surface in contact with the bone itself doesn't stimulate any positive reaction at the interface (f. e. osteoinduction or osteoconduction), because the polymer implant is not able to promote the bone healing. A Long-term aim of this study is therefore to produce new ceramic / polymer composite bioresorbable devices for internal fracture fixation of loaded bony fragments in maxillo-facial surgery with higher bioactive surface, in order to stimulate the bone repairing process and to provide at the same time an optimized stress transfer between implant and healed bones.

EXPERIMENTAL PROCEDURES

β-TCP raw powders (BK Giulini Chemie GmbH) were processed to be spray-dried in a co-current flow spray-drier in order to get suitable granulates for the Atmospheric Plasma Spraying (APS) technique. The aim is to obtain spherical granulates with a monomodal and narrow particle size distribution having a mean particle size D_{50} in the range 30-120 μm. These characteristics are requested to get the best spraying conditions, i.e. uniform melting of the powder in plasma stream, better flowability and constant feeding by means of conventional powder feeders. The manufacturing process and powders characterization have been reported in previous papers [10, 11]. The powders particle sizes are given in table I.

Table I. particle sizes of spray dried powders

Powder	P1	P2	P3	P4
D_{10} (μm)	12	33	32	134
D_{50} (μm)	82	119	115	300
D_{90} (μm)	194	340	234	597

Table II. Plasma spray parameter range used for spraying TCP

Substrate	PDLLA
Spraying distance (mm)	100 - 120
Gas flow Ar (l/min)	35 - 50
Gas flow H_2 (l/min)	6 - 8
Power (kW)	30 - 37

The coating experiments were carried out by means of a 6-axes computer controlled robot (Stäubli Unimation) equipped with a GTV plasma torch MF-P-1000 under the conditions listed in table II. An outstanding advantage of APS is the relatively low thermal load on the substrate during the coating process, so that it is possible to coat thermally sensitive substrates in an appropriate mechanical assembly with robot controlled plasma torches. Due to advanced cooling and APS parameters is it possible to control the thermal load on the substrate and it has been possible to coat temperature sensitive polymer devices. Compressed air was used for convective cooling. Thermal spraying on polymers may lead in the worst cases to complete or partial melting of the substrate, destroying it or at least modifying its properties. In particular in case of medical implants it is of paramount importance to avoid as much as possible surface modifications in order to maintain the original resorption properties of the polymer. The combination of fast torch movement and high cooling rate led to very satisfactory results (fig. 1). The partial melting of a sub-micron thick polymer layer at the interface substrate/coating is nevertheless unavoidable (fig. 2). Besides, this mechanism leads to mechanical adhesion of the coating to the substrate, which occurs as usual in thermal spraying by shrinkage of the molten particles at the impact with the substrate's surface.

The porosity measurements carried out combining light microscope and image point analysis showed very high porosity content. The light microscope shots evidence the presence of micro and macropores (fig. 2).

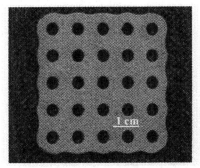

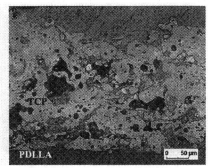

Fig. 1: Picture of a PDLLA implant plate coated with TCP by means of APS

Fig. 2: Light microscope shot of micro- pore structure in the TCP coating

The amount of porosity in the coatings, as well as the coatings' roughness, is influenced by the amount of unmolten particles, i.e. from the combined effect of spraying parameters, powders particle size and particle size distribution. P1 is actually the powder with the lower mean particle size and besides it was sprayed with higher plasma heat content, leading to uniform and complete melting of the particles in plasma and therefore to lower porosity and roughness. P2 and P3 on the other hand have a higher mean particle size and they were sprayed with lower plasma heat content. This led to incomplete melting and therefore to higher porosity and roughness. Roughness and porosity in relation to the powder type are given in table III. A scanning electron microscope analysis of the coatings surface (fig. 3) also showed the irregular and incomplete melting of the particles (splats).

Table III. Coatings porosity and roughness. Porosity determined by digital gray scale image analysis from light microscope cross sections.

Powder	Porosity [vol. %]	R_a [μm]	R_z [μm]	R_{max} [μm]
P1	12	9.1	54,3	60,3
P2	41	13,7	78,5	92,9
P3	31	13,9	79,2	95,8

A qualitative bending and shaping test was carried out at 70 °C. At this temperature it is still possible to easily modify the implant's shape without damaging or detaching of the ceramic coating (fig. 4), allowing to match the implant to the bone morphology. The mechanism of this behavior might be formation of microcracks in the coating, which allows the coating itself to be bended and shaped together with the substrates. Being the coatings highly porous, the microcracks are frequently stopped and are not able to run all through the coating, avoiding this way delamination and/or failure.

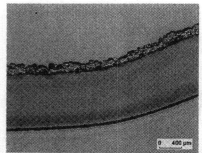

Fig. 3. SEM of the TCP coating's surface.

Fig. 4. Composite PDLLA/TCP after manual shaping and bending at 70 °C

Atmospheric plasma spraying is a high energetic process which involves high temperatures to melt any kind of spraying powder material. However, the melting process involves phase transformations which usually are not reversible. After being molten in the plasma stream, the particles impact on the substrate where they rapidly cool down (quenching). In particular regarding to TCP coatings is it of crucial importance to know whether α- or β-phase is present, or eventually a combination of the two, because of their different solubilities in the human body. The β-TCP particles in plasma undergo the phase transition β to α, which occurs at temperatures around 1120 °C. This transition may occur either completely or only partially, depending upon plasma parameters and powder properties. By means of XRD it is possible to determine the phase composition of each coating.

The qualitative analysis revealed two different types of coating compositions, i.e. coatings made of pure α-TCP phase and coatings where a mixture of α- and β-TCP is present (fig. 5). In the second case we assumed that only the external surface of each powder particle in plasma was molten, whereas the core did not reach the temperature of phase transition and therefore remained in the original state.

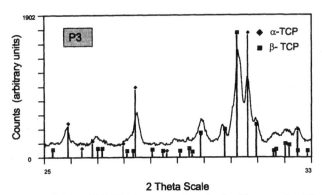

Fig. 5. XRD pattern of α- and β-TCP mixture coating obtained by spraying P3.

A semi-quantitative analysis of the coatings phases is also possible by means of XRD, since the intensities of the diffraction lines due to one phase of a mixture depend on the proportion of that phase in the mixture itself. In case of a two-phase mixture the following equation (2) permits to calculate the weight fraction w_β for the amount of β-TCP and consequently the amount of α-phase in each coating [12] can be estimated by:

$$\frac{I_\beta}{I_{\beta p}} = W_\beta . \quad (2)$$

Depending upon spraying parameters and powder properties, the calculated β-phase weight percent W_β comprised values between 11 and 26 %. It is evident that much higher β-TCP contents cannot be achieved, because the powder particles must be mostly molten in the plasma stream to get sufficient adhesion to the substrate. Non molten particles simply bounce off the substrate at the moment of the impact without creating any coating.

CONCLUSIONS

The manufacturing of composite materials which couple the properties of bioceramics and polymers is a promising way to solve some of the problems that these materials meet when they are implanted in the human body. To produce the TCP ceramic coatings on the polymer the thermal spraying process was chosen. This process requires appropriate powders. The powders were produced via spray-drying of aqueous slurries and optimized for the requirements of thermal spraying with automatic powder-feed equipment. Four different slurries were produced and subsequently spray-dried. Three of four spray-dried powders possess the required properties, i.e. monomodal particle size distribution, mean particle size in the range of 80 - 120 μm, good flowability and round shaped particles. The spray drying conditions now can be adapted from the pilot plant data to small series production. The as-sprayed TCP coatings present a much higher bioactive surface in relation to normal PDLLA implants. This is certified by the high roughness values and also by the high porosity (12.6 - 41.7 vol. %), with pore sizes ranging from few μm up to 100 μm and more (light microscope analysis). The coatings were also analysed by means of XRD in order to determine phase composition and crystallinity. Depending on the combination of spraying parameters and powder type, the coatings are made of pure α-TCP or of a mixture of β- and α-phase. The β-phase content ranges between 11 and 26 %. The new composite TCP/PDLLA maintains one of the most important characteristics of polymer implants,

15

i.e. the possibility to heat it up over the glass transition temperature in order to successfully bend it and shape it. This way the surgical implant can be adapted to the various shapes of the fractured bones.

Future work in the manufacturing process includes a complete mechanical characterization of the composite. An analysis of residual stresses must be performed, in order to create thermally sprayed coatings with residual stresses as low as possible. Residual stresses are induced by the mechanical impact of the particles on the substrate and also by the quenching mechanism, and they are present both in the coatings and in the substrate. In particular residual stresses in the substrate should be reduced to the minimum, in order to avoid a possible worsening of the implants mechanical properties. A further step will include testing of implants both *in vitro* and *in vivo*, in order to verify the advantages given by the ceramic coating.

REFERENCES

[1]de Groot K., Koch B., Wolke J.G.C., (1990): "X-Ray Diffraction Studies on Plasma-Sprayed Calcium Phosphate-Coated Implants", Journal of Biomedical Materials Research, Vol. 24, 655-667

[2]Hench L.L., (1998): "Bioceramics", J. Am. Ceram. Soc., 81 [7] 1705-28

[3]Bessho K, Iizuka T, Murakami K, (1997): "A bioabsorbable poly-L-lactide miniplate and screw system for osteosynthesis in oral and maxillofacial surgery", J Oral Maxillofac Surg 55: 941-945

[4]Bos R, Rozema F, Boering G, (1990): "Bioresorbable osteosynthesis in maxillofacial surgery", Oral and Maxillofacial Clinics of North America 2: 745-750

[5]Wintermantel E., Ha S.W., (1996): "Biokompatible Werkstoffe und Bauweisen. Implantate für Medizine und Umwelt", ed. Springer

[6]LeGeros J.P., LeGeros R.Z., (1993): "Dense Hydroxyapatite", in "An Introduction to Bioceramics", Advanced Series in Ceramics – Vol. 1, Editors Hench and Wilson, World Scientific

[7]Epple M., Eufinger H., Rasche C., Schiller C., Weihe S., Wehmöller M., (2001): "Ein optimierter biodegradierbarer Werkstoff für die Behandlung grossflächiger Schädeldefekte", Biomedizinische Technik, Band 46, Ergänzungsband 1

[8]Beckmann F., Epple M., Eufinger H., Rasche C., Schiller C., Weihe S., Wehmöller M., (2003): "Geometrically structured implants for cranial reconstruction made of biodegradable polyesters and calcium phosphate/calcium carbonate", Biomaterials, article in press

[9]Adam C, Hoffman J, Troitzsch D, Zerfowski M, Reinert S, (2003): "Bioresorbable polymer implants in maxillofacial trauma surgery", Eur Surg Res 35: 312-313

[10]Baccalaro, M , v. Niessen, K., Gadow, R.: "Manufacturing of Thermally Sprayed Tricalcium Phosphate (TCP) Coatings for Biomedical Applications". In: *Abstracts of 8th European Interregional Conference on Ceramics, CIEC8*, 03.-05. September 2002, Lyon

[11]Baccalaro, M.; Gadow, R.; v. Niessen, K.: "Manufacturing of thermally sprayed Tricalcium Phosphate TCP Coatings for Biomedical Applications". Symposium 2 on Bioceramics: Materials and Applications: a Symposium to honor Larry Hench, 105th American Ceramic Society Annual Meeting, April 2003, Nashville, Tenn., USA. AcerS Ceramic Transaction Vol. 147, ISBN: 1-57498-202-8

[12]Cullity B.D., (1978): "Elements of X-Ray Diffraction", 2nd Ed., Addison-Wesley ed., London

SYNTHESIS AND SINTERING STUDIES OF NANOCRYSTALLINE HYDROXYAPATITE POWDERS DOPED WITH MAGNESIUM AND ZINC

Himesh Bhatt
Department of Mechanical, Materials and Aerospace Engineering,
University of Central Florida
P.O. Box 162450
Orlando, FL-32816-2450

Samar J. Kalita
Department of Mechanical, Materials and Aerospace Engineering,
University of Central Florida
P.O. Box 162450
Orlando, FL-32816-2450

ABSTRACT
    In this research, we have synthesized nanocrystalline hydroxyapatite ($Ca_{10}(PO_4)_6(OH)_2$, Hap) powders doped with magnesium and zinc using the water-based sol-gel technique and characterized them. Calcium nitrate and triethyl phosphite were used as starting materials. These chemicals were dissolved in distilled water, separately, under vigorous stirring. As-prepared calcium nitrate sol was added drop wise into the hydrolyzed phosphite sol and then aged and dried. Dried gel was then crushed into fine white powders with the help of mortar and pestle and a measured amount of magnesium oxide and zinc oxide powders were added to the crushed amorphous powders, separately. Calcination was carried out at 250-500°C. Morphology of the powders was determined using transmission electron microscopy. TEM results revealed that the particle size diameter of powders were in the range of 5-10 nm. Phase analyses were carried out using powder X-ray diffraction technique. As-synthesized powders were also pressed uniaxially in a steel mold to prepare dense ceramic structures. These green structures were sintered at 1300°C for 6 h in a muffle furnace for densification. Highest sintered density of 3.29 g/cc was measured for magnesium-doped powder.

INTRODUCTION
    Hydroxyapatite ($Ca_{10}(PO_4)_6(OH)_2$, Hap) bioceramic has been most widely studied for bone engineering applications because of its excellent chemical stability and compositional similarities with the bone mineral.[1] However, Hap exhibits low mechanical strength and poor crack growth resistance, which limit its applications to non-load bearing applications for example as coatings or powders. To make Hap useful as bone grafts, it is critical to improve its mechanical strength and toughness. Some of the promising approaches that have been used to improve the mechanical performance of Hap are by doping with different sintering additives[2, 3], by introducing a glassy phase[4, 5], and by controlling important characteristic features of the powders such as particle size and shape, particle distribution and agglomeration.[6]
    Nanostructured ceramics offer unique advantages because of their high surface areas to volume ratio and unusual chemical synergistic effects. Objective of this research-project was to explore the arena of nanotechnology to improve mechanical performance of Hap ceramics. Accordingly, we focused our research on the synthesis of nanoscale Hap powder and then,

doped this powder with different metal ions (magnesium and zinc) at the molecular level to enhance mechanical performance. These metal ions are known to be present in the bone mineral and are believed to play a vital role in overall performance of bone.

A number of different techniques have been developed to synthesize HAp powder, which include sol-gel process,[7-10] solid-state reaction,[11] micro emulsion synthesis,[12] and hydrothermal reaction method.[13] Sol-gel processing is a favored technique due to its low synthesis temperature, homogenous composition, and high product purity. Sarig and Kahana highlighted the importance and advantages of nano crystalline HAp and synthesized powders with 300 nm edges.[14] Synthesis of nano-HAp via others routes has been also been attempted.[15] However, in most of these works, a long period of the sol preparation time, 24 h or longer, is commonly reported. In addition, preparation of single-phase HAp powders has been a concern. In this work, we report the synthesis of stoichiometric; nano crystalline, single-phase HAp powders within a short time frame (sol preparation time of 16 h) at a considerably low temperature range and then doped it with magnesium and zinc.

EXPERIMENTAL

A 0.025 mol of triethyl phosphite (Fisher, USA) was dissolved in a fixed amount of distilled water (the molar ratio of water to phosphite is fixed at 8) in a nalgene bottle under vigorous stirring. A stoichiometric amount, 0.045 mol of calcium nitrate (Fisher, USA) dissolved first in 25 ml of distilled water, was added drop wise into the hydrolyzed phosphite sol. The mixed sol solution was then continuously agitated for additional 4 min and kept static (aging) at 50°C. This aged sol was then subjected to thermal treatment at 85°C for 20h until a white and dried gel was obtained. The dried gel was ground into powder using a mortar and a pestle. Further, small quantities of magnesium oxide and zinc oxides were introduced into the ground powder, separately, and homogeneously mixed. These doped gels were further calcined at 250°C for 3 h, 350°C for 3 h and, 500°C for 15 min, at a constant heating rate of 15°C/min, followed by furnace cool. Phase analysis of the amorphous gel and the calcined powders was done using X-ray powder diffraction analysis (XRD) with Ni-filtered CuKα radiation (Rigaku Corp.; 40 kV, 40 mA). Morphology and powder particle-size were determined using transmission electron microscopy (HR-TEM, Technai-Phillips). As-synthesized nano-HAp powders were uniaxially compressed in a steel mold having an internal diameter of 10 mm at a pressure of 37.5 MPa. The green specimens were then sintered at a temperature range of 1100-1300°C for 6 h in a muffle furnace, in atmospheric conditions.

RESULTS AND DISCUSSION

Phase analyses

X-ray diffraction traces were obtained for the amorphous dried gel and the doped powders calcined at 250°C for 2 h, 350°C for 2 h and 500°C for 15 min, separately. Fig. 1a presents the XRD traces of the dried gel and the powders calcined at 250°C and 350°C. This figure shows that the dried gel exhibits highly amorphous characteristics. The powders calcined at 250°C and 350°C exhibits two broad peaks at about 30.5° with a considerable amount of amorphous phase. XRD traces of the powders calcined at 400°C and 500°C are presented in Fig. 1b. These traces showed that the apatite phase first appeared at 400°C and the HAp content increased with increase in calcination temperature. Powders calcined at 500°C showed the maximum intensity peaks corresponding to the planes (002), (210), (211), (202), (220), (310), (222), and (213).

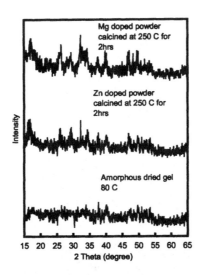

Fig. 1a XRD patterns of dried gel and, magnesium and zinc doped powders calcined at 250°C for 2 h.

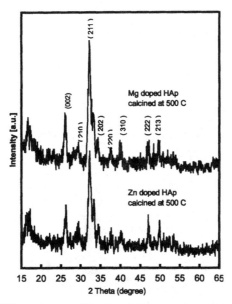

Fig. 1b XRD patterns of Magnesium and Zinc doped HAp powders calcined powders at 500°C.

19

Powder characterization

Transmission electron microscopy (TEM) was used to analyze the morphology of the as-synthesized doped powders calcined at 500°C for 15 min. Powders calcined at lower temperatures were not analyzed for their particle size. Results of TEM examinations are presented in Fig 2a and b respectively. It is evident from fig. 2(a) that the calcination of Mg doped powders at 500°C resulted into the agglomeration of nano HAp powders having an effective average particle size of 2-5 nm in diameter. Fig. 2(b) reveals that the calcination of Zn doped powders at 500°C resulted into agglomerated nano HAp powders having particle size in the range of 20-50 nm in diameter. Similar results were obtained when powder synthesis and characterization experiments were repeated.

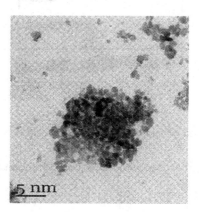

Fig. 2a TEM micrograph of Mg doped nano HAp powder calcined at 500°C for 15 min, (1.0 wt% of Mg)

Fig. 2b TEM micrograph of Zn doped nano HAp powder calcined at 500°C for 15 min, (1.0 wt% of Zn).

Densification studies

The green ceramic compacts prepared via uniaxial pressing were subjected to pressure-less sintering in a muffle furnace at 1250°C and 1300°C for 6 h, separately. Four green specimens of each of the compositions (containing 1.0, 2.5 or 4.0 wt% of additives) were used to study the effect of particle size, presence of metal ion dopants and, sintering temperature on the densification process. Bulk density of the green and the sintered structures were measured. Average green and sintered densities of each of these compositions are presented in Fig. 3(a) and Fig. 3(b). These figures show variation of density as a function of amount of additives. It is clear from Fig. 3(a) sintering at 1250°C helped in densification of HAp and HAp doped with Mg and Zn. A maximum sintered density of 3.12 g/cc was obtained for the Mg doped nano HAp (1.0 wt.%). Fig. 3(b) shows that sintering at 1300°C has significantly improved densification of HAp and HAp doped with Mg and Zn. A maximum sintered density of 3.29 g/cc was obtained for the Mg doped nano HAp (1.0 wt.%). Fig. 3(b) also shows that compared to pure HAp, all composition of Mg and Zn doped HAp showed higher densification at 1300°C.

Hardness testing

A Vickers hardness tester was used to determine hardness of Mg and Zn doped dense ceramic nano compacts with 1.0 wt.% additives, sintered at 1300°C, for 6 hours. Fig. 4 shows Vickers hardness, expressed in percentages, recorded for these compositions as a function of sintered density at 1300°C. It is clear from this figure that with improved densification, the hardness increased. The highest hardness was recorded for nano HAp structure doped with

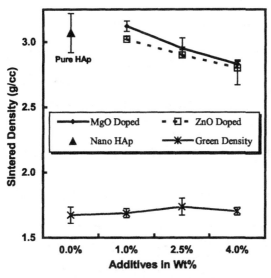

Fig. 3(a). Influence of magnesium and zinc doping on densification of nano HAp. Sintering was done at 1250°C for 6 h.

21

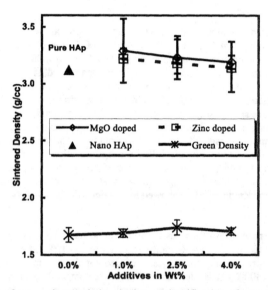

Fig. 3(b). Influence of magnesium and zinc doping on densification of nano HAp. Sintering was done at 1300°C for 6 h.

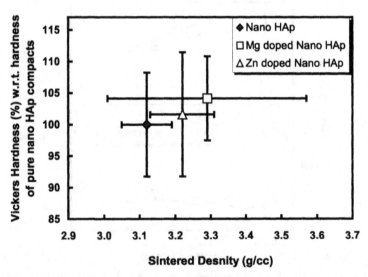

Fig. 4 Vickers hardness (in percentage) as a function of sintered density at 1300°C.

1.0 wt.% of Mg. Zn doped nano HAp showed marginal increase in hardness value over the pure HAp compacts. It is evident from the figure, that there is a considerable rise; approximately 5%, in hardness for Mg doped nano HAp with 1.0 wt.% doping compared to pure nano HAp.

Compression testing
 Sintered ceramic structures were tested for their compression strength using a fully automated screw driven uniaxial tensile tester from Instron. Specimens were prepared for pure nano-HAp and nano-Hap doped with 1.0 wt% of Mg and Zn. These specimens had diameter to height ratio of 1:1.5. These tests were carried out with a constant crosshead speed of 1.0 mm/min. Five specimens of each of these compositions were tested for their compression strength. Average strength of each of these compositions was calculated. Results of compression testing are shown in Table 1. It is clear from these results that presence of Mg and Zn dopants significantly increased compression strength of nano-HAp. Three-fold increase in strength was observed in nano-HAp doped with 1.0 wt% of Mg. These results show that homogenous doping with metal ions such as Mg and Zn could enhance mechanical performance of nano-HAp. Structures prepared via high pressure pressing routes such as cold–isostatic pressing and hot – isostatic pressing will possibly exhibit better mechanical properties.

Table 1 Failure strengths of nano HAp, and Mg and Zn doped nano HAp structures under uniaxial compressive loading.

Type	Compression Strength (MPa)
Pure NanoHAp	30.1 (±6.4)
Mg doped Nano HAp	91.2 (±5.3)
Zn doped Nano HAp	68.9 (±6.8)

CONCLUSION
 The phenomenon of sol to gel and gel to ceramic during the synthesis of magnesium and zinc doped nano hydroxyapatite powder, through the water-based technique, was investigated using high resolution TEM and X ray diffractometer. It was observed that the calcined Mg doped powder was an assembly of fine nano-sized particles of size ranging from 2-5 nm in diameter. Powder particle-size of Zn doped powder was found to be 20-50 nm in diameter.
 Powder XRD analysis showed the occurrence of HAp phase shortly after calcination at 400°C. The densification studies reveal that the density of nanostructured HAp increased with the increase in sintering temperature. A maximum sintered density of 3.29 g/cc was achieved for nano HAp doped with Mg, sintered at 1300°C. Presence of dopants improved densification, hardness and compression strength of nano-HAp.
 This study concludes that the doped nano-structures of HAp ceramics has the potential to overcome some of the limitations related to mechanical performance. However, detailed *in vitro* and *in vivo* studies are necessary to make these ceramics useful as bone grafts in clinics.

REFERENCES

[1]L.L. Hench, "Bioceramics: from concept to clinic" *J. Am. Ceram. Soc.*, **74**, 1487–510 (1991).

[2]W. Suchanek, M. Yashima, and M. Yoshimura, "*Hydroxyapatite ceramics with selected sintering additives*" Biomaterials, **18**, 923-933 (1997).

[3]S. J. Kalita, D. Rokusek, S. Bose, H. L. Hosick, and A. Bandyopadhyay, "Effects of MgO-CaO-P₂O₅-Na₂O-based additives on mechanical and biological properties of hydroxyapatite," *J. of Biomed. Mater. Res. Part A*, **71A**, 35-4 (2004).

[4]J. D. Santos, P. L. Silva, J. C. Knowles, S. Talal, and F.J. Monteiro, "Reinforcement of Hydroxyapatite by Adding P₂O₅-CaO Glasses with Na₂O, K₂O and MgO". *J. Mat. Sci. Mat. in Med.*, **7**,187-9 (1996).

[5]J. C. Knowles, S. Talal, and J. D. Santos, "Sintering effects in a glass reinforced hydroxyapatite," *Biomaterials*, **17 [14]**, 1437-42 (1996).

[6]R. Kumar, K.H. Prakash, K. Yennie, P. Cheang, and K.A. Khor, "Synthesis and characterization hydroxyapatite nano rods/whiskers," *Key Engineering Material*, **284-286**, 59-62 (2005).

[7]D.M. Liu, Q. Yang, T. Troczynski, and W.J. Tseng, "Structural evolution of sol-gel derived hydroxyapatite,"*Biomaterials*, **23**, 1679 (2002).

[8]T.K. Anee, M. Ashok, M. Palanichamy, and S.N. Kalkura, "A novel technique to synthesize hydroxyapatite at low temperature,"*Mater. Chem. Phys*, **80**, 725 (2003).

[9]R.H. Takahashi, M. Yashima, M. Kakihana, and M. Yoshimura "Synthesis of stoichiometric hydroxyapatite by a sol-gel route from the aqueous solution of citric and phosphoneacetic acids," *Eur. J. Solid State Inorg. Chem.*, **32**, 829-35 (1995).

[10]P. Layrolle, A. Ito, and T.S. Teteishi, "Sol-gel synthesis of amorphous calcium phosphate and sintering into microporous hydroxyapatite bioceramics," *Journal of American Ceramic Soc.*, **81**, 1421 (1998).

[11]K.C.B. Yeong, J. Wang, and S.C. Ng, "Mechanochemical synthesis of nanocrystalline hydroxyapatite from CaO and CaHPO₄," *Biomaterials*, **22**, 2705 (2001).

[12]G.K. Lim, J. Wang, S.C. Ng, C.H. Chew, and L.M. Ganl, "Processing of hydroxyapatite via microemulsion and emulsion routes," *Biomaterials*, **18** 1433-1439 (1997)

[13]H.S. Liu, T.S. Chin, L.S. Lai, S.Y. Chiu, K.H. Chung, C.S. Chang, and M.T. Lui, "Hydroxyapatite synthesized by a simplified hydrothermal method," Ceramics International, **23** 19-25 (1997)

[14]S. Sarig, and F. Kahana, "Rapid formation of nanocrystalline apatite," *Journal of Crystal Growth*", **237**, 55 (2002).

[15]H.K. Varma, S.N. Kalkura, and R. Sivakumar, "Synthesis and sintering of nanocrystalline hydroxyapatite powders,"*Ceram. International*, **24**, 467 (1998).

SEQUENCE SPECIFIC MORPHOLOGICAL CONTROL OVER THE FORMATION OF GERMANIUM OXIDE DURING PEPTIDE MEDIATED SYNTHESIS

Matthew B. Dickerson, Ye Cai, and Kenneth H. Sandhage
School of Materials Science and Engineering, Georgia Institute of Technology, Atlanta, GA, 30332, USA
Rajesh R. Naik
Materials and Manufacturing Directorate, Air Force Research Laboratory, Wright Patterson AFB, Ohio 45433, USA
Morley O. Stone
Defense Advanced Research Projects Agency, Arlington, VA 22203, USA

ABSTRACT
Peptides isolated from a phage library have been effective in promoting the precipitation of amorphous germanium oxide from a precursor solution. The morphologies of germania formed under the influence of these library-isolated peptides were shown to be distinct from the GeO_2 formed under similar conditions with homo poly-amino acids. The appearance of germanium oxide hollow spheres or core/shell nanostructures created under the influence of peptide Ge8 is clearly linked to the unique amphiphilic character of the peptide. The appearance of these unique structures represents the first time such morphologies have been seen in germania or created utilizing only an unmodified peptide. The ease of creating these interesting inorganic materials makes this system attractive for the possible fabrication of materials appropriate for catalysis, optical and controlled release applications.

INTRODUCTION
Recently, much attention has been paid to the preparation of oxide nanomaterials through biogenic or biomimetic approaches.[1-5] A major portion of the work conducted to date has been focused upon the use of proteins isolated from organisms or biological analogues (i.e., poly-l-lysine) for the formation of silica or calcium carbonate materials. Indeed, there have been over 150 studies in the last decade involving the formation of silica under the influence of poly-l-lysine. Similar investigations into the hydrolysis and condensation of germanium oxide by biological molecules are noticeably absent from the primary literature.

Although germanium oxide is of minimal biological relevance,[6] and there are no known germanium oxide analogs of the intricate hierarchically-organized silica structures produced by diatoms or sponges, this certainly does not mean that studies into the biomimetic formation of germania are irrelevant. Indeed, germania-based and GeO_2-containing glasses are important to the ceramic engineering community as they possess excellent infrared transmission and are key components of optical fibers.[7-10] The bottom up biomimetic formation of nanostructured germanium oxide glasses could also be a key technology in the development of a wide range of new sensors, display devices and waveguides for integrated optical systems. The enhanced solubility of germania over silica in water also makes this oxide interesting for controlled release applications in medicine and agriculture.[11-12]

An effective technique for the identification of proteins specific to a target surface or molecule is the phage displayed peptide screening method. This technique, popularly known as biopanning, has been used to isolate peptides that are specific to a variety of inorganic material surfaces, including III-V semiconductors, noble metals, and a handful of oxide ceramics.[3,4, 13-15]

Several of these studies have also identified peptides of unique sequence which proved effective in controlling the morphology or crystal structure of the materials prepared under their influence.[4, 16] The objective of the present paper is to evaluate correlations between the amino acid sequence of peptides isolated through the phage-displayed library screening process, with the morphologies of germanium oxide structures produced under their influence.

PROCEDURE

Poly-amino acids (Sigma Aldrich, St. Louis, MO) were used as purchased without further purification or characterization. 5 mM stock solutions based upon the molecular weight of the amino acid monomer (not the polymer weight) of the various poly-amino acids were prepared by adding the appropriate mass of powdered poly-amino acid to 500 μl of nanopure water. A detailed description of the phage-displayed peptide screening method has been previously provided[17]. Stock solutions of the phage-displayed library isolated peptides were created using an averaged amino acid molecular weight which was derived by dividing the peptide molecular weight by 12. Germanium oxide precipitation was conducted by diluting 10 μl of peptide stock solution into 40 μl of anhydrous methanol (99.9%, Alfa Aesar, Ward Hill, Ma). 50 μl of an anhydrous methanol-based solution containing tetramethoxygermanium (TMOG) (99.999%, Alfa Aesar, Ward Hill, Ma) at a concentration of 0.27 M added to the diluted peptide solution and mixed by inversion several times. The resulting precipitates were concentrated by centrifugation and repeatedly washed with anhydrous methanol. The amount of germania precipitated from the alkoxide solution (i.e., the germania precipitation activity) in the presence of a given peptide was determined by adapting the β-silicomolybdate method described by Iler.[18] The germanium oxide precipitate was first dissolved in boiling 1 M NaOH for 20 minutes. The reaction of the molybdic acid with hydrolyzed germania gave a yellow product with an absorption maximum at 410 nm. The absorption value of the sample was found to be linearly related to the concentration of germania in the sample within the investigated limits.

RESULTS AND DISCUSSION

In order to better assess the unique character of peptides isolated through the phage-displayed library screening technique, the germanium oxide-forming nature of several commercially available homo poly-amino acids were investigated. The eight poly-amino acid peptides utilized in this study were selected to encompass the side group character spectrum of the 20 common biologically produced amino acids.

Biomimetic Formation of GeO₂ with Poly-Amino Acids

Addition of the germanium oxide precursor solution to either of the anionic amino acid residues, poly-aspartic acid or glutamic acid, failed to produce any detectable precipitation product (Fig. 1). The lone hydrophobic amino acid representative, poly-alanine, also failed to produce germanium oxide from the precursor solution. Inspection of Fig. 1 reveals that the poly-amino acids composed of residues capable of participating in hydrogen bonding interactions (i.e., threonine and asparagine) exhibit modest germania formation activity. Of poly-threonine and poly-asparagine, the hydroxyl-rich poly-threonine possesses greater precipitation activity. The results of the interaction of poly-alanine, asparagine, threonine, glutamic acid and aspartic acid in the formation of germanium oxide correlate well with their previously observed interactions or lack of interaction with chemically-similar silica precursors.[19-21] As expected from previous biomimetic SiO₂ research,[19-21] the cationic poly-amino acids exposed to germania precursor molecules exhibited high levels of germanium oxide precipitation. Unfortunately, the level of standard deviation in the amount of germania produced by poly-lysine, poly-histidine, and poly-arginine as well as the uncertainty associated with preparing solutions based upon dry peptide weight (which includes salts) makes the assignment of the most efficient germania-forming poly peptide from this group unclear.

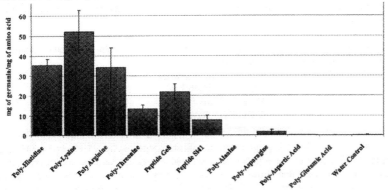

Figure 1: Precipitation activity of several poly-amino acids, peptide Ge8, and peptide Si41.

Observations of the poly-amino acid precipitated germanium oxide product were conducted by SEM and TEM, the results of which are presented in Figures 2a and 2b, respectively. Although the chemical character, polymer length, and the germanium oxide precipitation activities of the studied poly-amino acids differ substantially, the morphology of their GeO_2 precipitates is remarkably similar. Such germania structures appear as necked particles approximately 50 nm in size and exhibit an extremely rough surface (Fig. 2a). Energy dispersive spectroscopy conducted simultaneously with SEM observations confirmed the presence of germanium and oxygen and the absence of other elements in the precipitates. Higher magnification interrogation of these poly-amino acid induced germanium oxide precipitates (Fig. 2b) revealed that these structures were composed of a cottony mass of amorphous material. Electron diffraction of these precipitates produced only a diffuse ring pattern indicative of an amorphous material (Fig. 2b inset).

Figure 2: a) Secondary electron and b) high resolution transmission electron micrograph of typical germanium oxide precipitate morphologies produced by poly-K, R, H, T, and N.

27

Library-Isolated Peptide Induced Germania Precipitation

Twenty one individual germanium binding peptides were isolated from the phage displayed library screening work conducted in this study. A complete listing of the primary structure of these 12 amino acid peptides has been previously reported.[17] Three peptides (Ge8, Ge34, and Ge2) which are representative of isoelectric character of these 21 unique peptide sequences were selected for further study. The amino acid sequence and isoelectric point of these three peptides are listed in Table I.

Table I : Amino acid sequences and calculated isoelectric points of three germania-binding peptides isolated from a phage displayed peptide library.

Peptide	Amino Acid Sequence	pI[a]
Ge2	TSLYTDRPSTPL	5.50
Ge8	SLKMPHWPHLLP	8.51
Ge34	TGHQSPGAYAAH	6.61

[a]pI calculated using pI/mass program at www.expasy.ch,

The addition of a TMOG solution into the diluted peptides Ge2, Ge8, and Ge34 resulted in an immediate clouding of the solution and yielded a product capable of concentration by centrifugation. In contrast, the addition of an equivalent volume of water (i.e., no peptides) did not yield any detectable solid materials. The morphologies of the peptide-generated precipitates were investigated by scanning and transmission electron microscopy techniques, the results of which are presented in Figs. 3-5.

Figure 3: (left) Secondary electron micrograph of the germania precipitated under the influence of peptide Ge2.

The interaction of peptide Ge2 (which possessed the lowest pI and germanium oxide precipitation activity[17] of the library-isolated peptides) with TMOG yielded a heavily fused mass of germanium oxide particles approximately 500 nm in size, as evident in Figure 3. The morphology of GeO_2 precipitated with Ge34 (Fig. 4a) is similar to that formed by Ge2, yet more refined. Again, the germania precipitate is typified as a interconnected mass of particles. The high resolution TEM micrograph of germania precipitate prepared with peptide Ge34 (Fig. 4b) correlates well with the previous SEM observations of this sample. Distinct particles are indiscernible, and instead, a highly branched network is visible. The fused appearance of the Ge2 and Ge34 precipitated germania contrasts quite heavily with that of the germanium oxide generated by the hydrophilic poly-amino acids discussed previously. Particle fusion and the transition of discrete silica particles into films have been previously observed to occur with the addition of molecules with hydroxyl moieties (e.g., ethylene glycol or saccharides) to poly-lysine solutions prior to precursor exposure.[22] Although the library isolated Ge2 and Ge34 peptides

28

contain hydroxyl moieties, their lone influence over surface morphology is somewhat perplexing as the germania produced by the hydroxyl rich peptide poly-threonine is identical to that produced by poly-amino acids lacking OH groups (e.g., lysine or arginine). Attempts to mimic the chemistry of the library-isolated peptides through the use of a mixture of poly threonine and poly-lysine or poly- histidine failed to produce germania that was morphologically distinct from that produced under single poly-amino acid conditions (Fig. 2a).

Figure 4: a) Secondary electron and b) high resolution transmission electron microgaphs of germanium oxide formed from a TMOG solution by peptide Ge34.

The morphology of germanium oxide produced under the influence of peptide Ge8 is quite different from that of the GeO_2 formed by either homo poly-amino acids, peptide Ge2, or peptide Ge34. Assessment of the scanning electron micrograph presented in Figure 5 reveals the presence of round germanium oxide particles 100-600 nm in size. The particles appear to have a bimodal size distribution, with small particles having diameters clustered around 150 nm and larger particles that are approximately 550 nm in diameter. The transmission electron microscopy of the smaller group of particles (Fig. 6a) confirms that although these particles are somewhat fused together, they are recognizably discrete, with sizes ranging down to 30 nm. As with the previously described peptide produced germania, electron diffraction indicated a lack of crystallinity in the germanium oxide particles produced by Ge8. Evaluation of the larger particle class of Ge8 generated germania by TEM revealed either a hollow sphere or core/shell architecture. The SEM-observable shells of these structures possess a range of diameters, with sizes of 550 nm being typical. Encapsulated within a large portion of these shells are solid cores of amorphous germanium oxide which possess diameters on the order of 350 nm. These observations represent the first report of such hollow sphere or core/shell germania structures.

The formation of these distinct Ge8 created structures is most likely occurring through the aggregation or self-assembly of the Ge8 peptide. Mechanistically, such peptide organization is likely to occur to minimize the exposure of the hydrophobic PLL tail of the Ge8 peptide. Leucine-leucine interaction is a well known phenomenon in the structural biochemistry and is often seen in the packing of two secondary structural elements together, such as helices or sheets.[23] The much larger precipitate size of even the smaller class of Ge8 induced germanium oxide particles is evidence of such a peptide agglomeration event. One might expect the size of the precipitate to scale with the size of the precipitating scaffold peptide. However, the size of the poly-amino acids utilized in this study were 2 to 5 times larger than that of peptide Ge8, yet their precipitate size is 3 to 10 times smaller than that produced by Ge8.

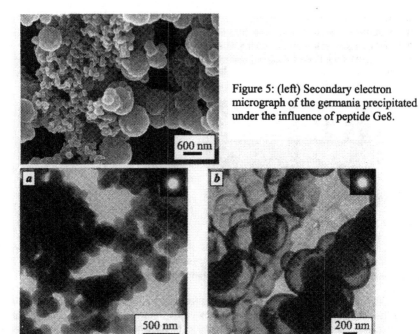

Figure 5: (left) Secondary electron micrograph of the germania precipitated under the influence of peptide Ge8.

600 nm

500 nm

200 nm

Figure 6: HRTEM microgaphs of the two morphologies of germanium oxide formed from a precursor solution by peptide Ge8.

The morphology and size of the Ge8 nanoparticles resembles the silica spheres precipitated *in vivo* by the R5 peptide isolated from the diatom *Cylindrotheca fusiformis*[1]. Recent studies[1,2] have shown that the R5 peptide self organizes and agglomerates through interpeptide bridging interactions which exploit phosphate ions in the buffer solution. In the present study phosphate ions are absent, and instead the formation of a hydrophobic leucine core is likely to drive peptide organization. A similar result may have been previously indirectly observed in an R5 truncation study in which peptide fragments possessing the nonbiologically active RILL segment of the R5 peptide were found to have the greatest silica precipitation activity.[24]

Materials synthesis techniques utilizing micelle structures composed of amphiphilic molecules have been widely studied, especially for the formation of nanoparticles or hollow sphere structures.[25] Two examples of the formation of hollow silica spheres utilizing amphiphilic block copolymers or modified poly-lysine molecules have recently appeared in the literature.[26, 27] Similarly, the amphiphilic character of the Ge8 may lead to the formation of micelles or membrane structures where the hydrophilic hydroxyl and cationic side groups of the peptide are exposed to solution making them free to catalyze the formation of germania. Clearly, the gap between the inner germanium oxide core and shell wall was too great to have been generated by two sides of a classical double membrane like structure (i.e., the 75 nm distance between solid germania core and shell wall is well in excess of the length of two fully extended 12 residue peptides). Therefore, the distance between the shell and the core most likely do not represent materials formed on either side of the peptide membrane. Instead, the formation of a large 550 nm membrane encapsulating a smaller micelle, double membrane, or random mass of Ge8 peptides is most likely occurring.

The germania precipitation efficiency of peptide Ge8 appears to perform comparably with the cationic poly-amino acids investigated in this study. Interestingly, Ge8 slightly outperforms peptide Si41, a peptide isolated in biopanning against silica.[3] Peptide Si41 possesses a higher density of active hydroxyl and cationic residue side chains then does peptide Ge8. The increased germania formation activity level of Ge8 may correlate with the proposed self organization of the ampiphilic peptide, and indicates that specific sequence of the peptide is critical in obtaining maximum germanium oxide precipitation efficiency.

CONCLUSION

The morphology and germania precipitation activity of eight poly-amino acids, and three peptides isolated against germanium from a phage displayed peptide library were explored. Anionic and hydrophobic poly-amino acids proved ineffective in the precipitation of germania from a TMOG precursor. The amorphous germanium oxide materials precipitated under the influence of five hydrophilic poly-amino acids were observed to be morphologically identical. However, all germanium oxide produced by the library-isolated peptides was quite distinct from that created by commercially available homo poly-amino acids. Germanium oxide core/shell nanostructures were produced through the self organization of peptide Ge8. The formation of double membrane like architectures through interpeptide leucine interactions followed by surface condensation of germania is proposed as the mechanism responsible for the formation of such unique structures.

ACKNOWLEDGMENTS

This work has been funded by a grant provided by the US Air Force Office of Scientific Reserch under program managers Joan Fuller and Hugh DeLong.

REFERENCES

[1] N. Kroeger, S. Lorenz, E. Brunner, and M. Sumper, "Self-Assembly of Highly Phosphorylated Silaffins and Their Function in Biosilica Morphogenesis," *Science*, **298**, 584 (2002).

[2] M. Sumper, and N. Kroeger, "Silica Formation in Diatoms: the Function of Long-Chain Polyamins and Silafifins," *J. Mater. Chem.*, **14**, 2059, (2004).

[3] R. Naik, L. Brott, S. Clarson and M. Stone, "Silica-precipitating peptides isolated from a combinatorial phage display peptide library, "*J. Nanosci. Nanotech.*, **2(1)**, 95, (2002).

[4] R. Naik, S. Stringer, G. Agarwal, S. Jones, and M. Stone, "Biomimetic synthesis and patterning of silver nanoparticles," *Nature Materials*, **1(3)**, 169, (2002).

[5] S. Brown, M. Sarikaya, and E. Johson, "A Genetic Analysis of Crystal Grwoth," *J. Mol. Biol.*, **299**, 725, (2000).

[6] J. Frausto da Silva and R. Williams, The Biological Chemistry of the Elements, Oxford Press, New York, 2001.

[7] K. Kojima, K. Tsuchiya and N. Wada, "Sol-Gel Synthesis of Nd^{3+}-Doped GeO_2 Glasses and Their Optical Properties, "*J. Sol-Gel Sci. Technol.*, **19**, 511, (2000).

[8] C. Layne, W. Lowdermilk and M. Weber, "Multiphonon Relaxation of Rare-Earth Ions in Oxide Glasses," Phys. Rev B., **16(1)**, 10-20, (1977).

[9] S. Honkanen and S. Jiang, Rare-Earth Doped Devices II, International Society for Optical Engineering, Washington, DC, 1998

[10] G. Brusatin, M. Guglielmi and A. Marucci, "GeO$_2$-Based Sol-Gel Films," *J. Am. Ceram. Soc.*, **80**, 3139, (1997).

[11] M. Krishna and H. Hill, "Germanium Oxide Systems. III. Solubility of Germania in Water," *J. Am. Ceram. Soc.*, **48(2)**, 109, (1965).

[12] I. Takeda and K. Yoshie, US Patent Application 2000-638617, (2002).

[13] S. Whaley, D. English, E. Hu, P. Bararbara and A. Belcher, "Selection of Peptides with Semiconductor Binding Specificity for Directed Nanocrystal Assembly," *Nature*, **405**, 665, (2000).

[14] R. Naik, S. Jones, C. Murray, J. McAuliffe, R. Vaia, and M. Stone, "Peptide Templates for Nanoparticle Synthesis Derived from Polymerase Chain Reaction-Driven Phage Display," *Adv. Funct. Mater.* **14(1)**, 25, (2004).

[15] D. Gaskin, K. Starck, and E. Vulfson, "Identification of Inorganic Crystal-Specific Sequences using Phage Display Combinatorial Library of Short Peptides: A Feasibility Study," *Biotechnology Letters*, **22**, 1121, (2000).

[16] C. Flynn, C. Mao, A. Hayhurst, J. Williams, G. Georgiaou, B. Iverson, and A. Belcher, "Synthesis and Organization of Nanoscale II VI Semiconductor Materials Using Evolved Peptide Specificity and Viral Capsid Assembly," *J. Mater. Chem.*, **13**, 2414, (2003).

[17] M. Dickerson, R. Naik, M. Stone, Y. Cai, and K. Sandhage, "Identification of Peptides that Promote the Rapid Precipitation of Germania Nanoparticle Networks *via* us of a Peptide Display Library," *Chem. Commun.*, **15**, 1776, (2004).

[18] R. Iller, The Chemistry of Silica, Wiley, New York, 1979.

[19] D. Belton, G. Paine, S. Patwordhan, and C. Perry, "Towards an Understanding of (Bio)Silicification: the Role of Amino Acids and Lysine Oligomers in Silicification," *J. Mater. Chem.*, **14**, 2231, (2004).

[20] L. Sudheendra, and A. Raju, "Peptide-Induced Formation of Silica from Tetraethyl Orthosilicate at Near-Neutral pH," *Mater. Res. Bull.*, **37**, 151, (2002).

[21] T. Coradin, O. Durupthy, and J. Livage, "Interactions of Amino-Containing Peptides with Sodium Silicate and Colloidal Silica: A Biomimetic Approach of Silicification," *Langnuir*, **18**, 2331, (2002).

[22] E. Vrieling, T. Beelen, R. Van Santen, W. Gieskes, "Mesophases of (Bio)Polymer-Silica Particles Inspire a Model for Silica Biomineralization in Diatoms," *Angewandte Chemie*, **41(9)**, 1543, (2002).

[23] R. Garrett and C. Grishman, Biochemistry, Saunders College Publishing, New York, 1999.

[24] M. Knecht and D. Wright, "Functional Analysis of the Biomimetic Silica Precipitating Activity of the R5 Peptide from *Cylindrotheca fusiformis*," *Chem. Commun*, **24**, 3038, (2003).

[26] M. Adachi, T. Harada, and M. Harada, "Processes of Silica Network Structure Formation in Reverse Micellar Systems," *Langmuir*, **17(14)**, 4189, (2001).

[27] K. van Bommel, J. Jung, and S. Shinkai, "Poly(L-lysine) Aggregates as Templates for the Formation of Hollow Silica Sphere," *Adv. Mater.*, **13(19)**, 1472, (2001).

[28] E. Veriling, Q. Sun, T. Beelen, S. Hazellar, W. Gieskes, R. Van Santen, and N. Sommerdijk, "Controlled Silica Synthesis Inspired by Diatom Sicon Biomineralization," *J. Nanosci. Nanotech.*, **5**, 68, (2005).

SYNTHESIS OF NANO-SIZE HYDROXYAPATITE (HAp) POWDERS BY MECHANICAL ALLOYING

Soon Jik Hong, Himesh Bhatt, C. Suryanarayana, Samar J. Kalita
Department of Mechanical, Materials and Aerospace Engineering
University of Central Florida
Orlando, FL 32816-2450

ABSTRACT

Nano hydroxyapatite ($Ca_{10}(PO_4)_6(OH)_2$, HAp) powders were synthesized by solid-state reaction of $Ca(OH)_2$ and P_2O_5 mixtures in a high-energy SPEX 8000 shaker mill, using hardened steel vial and balls. The phase analysis was carried out using X-ray powder diffraction technique. Transformation of $Ca(OH)_2$ and P_2O_5 mixture to HAp phase was first observed after 1 h of milling. The powder milled for 3 h showed prominently the presence of HAp phase. TEM analysis revealed that as-synthesized HAp powder was in the range of 20-60 nm. Measured quantities of synthesized nano-powders were pressed uniaxially in a steel mold to prepare dense ceramic structures for densification studies. These green structures were subjected to sintering studies at 1300 °C for 6 h when the highest sintered density of 3.17 g/cc was achieved.

INTRODUCTION

Synthetic hydroxyapatite (HAp) is a representative material for bone substitutes because of its excellent biocompatibility and compositional similarity with the inorganic phase of bone. A number of different processes have been developed and used to synthesize HAp powders for commercial applications.[1-6] Most of these processes can be broadly classified into two groups - wet methods[1-3] and dry methods[4-6]. The advantages of wet methods are that the by-product is always water and the probability of contamination during processing is low. However, the composition of the resulting product is greatly affected by even a slight difference in the reaction conditions and the time needed for obtaining hydroxyapatite of stoichiometric composition is relatively longer, which is inconceivable for industrial production. In addition, handling of the precursor materials and the apparatus, in wet methods, is complicated which results in poor reproducibility and high processing costs. Therefore, when producing high crystalline HAp powders in large volumes, the dry processes are the preferred methods because of their high reproducibility and low processing cost in spite of the risk of contamination during milling.

Mechanical alloying (MA) is a solid-state powder processing method that involves repeated cold welding, fracturing, and rewelding of powder particles in a high-energy ball mill.[7,8] It has been shown that MA can synthesize many types of material such as oxide-dispersion strengthened (ODS) superalloys, intermetallics, amorphous alloys, nanostructured materials, and a variety of other non-equilibrium phases.[8-10] It would be of our interest to conduct mechanical milling studies on stoichiometric powder mixture of calcium hydroxide [$Ca(OH)_2$]and phosphorous pentoxide [P_2O_5] to evaluate the mechanochemical solid-phase reaction for the synthesis of nano-grained hydroxyapatite ($Ca_{10}(PO_4)_6(OH)_2$, HAp). MA needs precise control for the preparation of HAp. The composition and properties of the final product are strongly influenced by parameters such as milling time, ball to powder ratio, and atmosphere.[9,10]

In the present work, the mechanical alloying of HAp from its constituent powder mixtures viz., calcium hydroxide [$Ca(OH)_2$] and phosphorous pentoxide [P_2O_5] was attempted.

The synthesis of HAp would be mainly achieved by a solid-state reaction between the solids activated either at the surface or in the bulk by intensive mechanical alloying. The present work, therefore, describes the synthesis and characterization of nanocrystalline HAp powder via mechanical solid-state reaction.

EXPERIMENTAL PROCEDURE

Stoichiometric amounts of elemental powders of $Ca(OH)_2$ (Alfa Aesar, 95 % pure) and P_2O_5 (Alfa Aesar, 96.5 % pure) were mixed together to obtain the desired phase of $Ca_{10}(PO_4)_6(OH)_2$ (hydroxyapatite: HAp). Mechanical alloying was conducted in a SPEX 8000 mixer mill using a hardened steel vial with 6 mm diameter hardened steel balls. The main channels by means of which oxygen gets into the mechanical alloying product are surface oxides and oxygen trapped from the atmosphere during milling. To reduce contamination associated with oxygen and oxides, loading and unloading of the powders was always conducted in a glove box, which was constantly maintained under a protective argon atmosphere.

For each experiment, 10 g of the blended elemental powder mixture ($Ca(OH)_2$ and P_2O_5) and 100 g of the hardened steel balls were loaded into the milling container. The ball-to-powder weight ratio was maintained at 10:1 during milling.

Scanning electron microscopy (SEM) was employed to determine the size and shape of the powders synthesized. The powder particle size at different milling times was measured from TEM micrographs. X-ray diffraction (XRD) patterns were recorded at different stages of milling with the help of a Rigaku diffractometer using CuKα radiation (λ =0.1542 nm) at 35 kV and 35 mA settings. The XRD patterns were recorded in the 2θ range of 10 to 70 degrees, with a step size of 0.02 degrees and step duration of 0.5 s.

Powder mixtures of $Ca(OH)_2$ and P_2O_5 after different milling times (0 min, 30min, 60 min and 180 min) were compacted uniaxially in a steel mold having an internal diameter of 10 mm at a pressure of 37.5 MPa. The green specimens were then sintered at 1300 °C for 6 h in a muffle furnace, in air. As-sintered specimens were evaluated for their bulk density to understand the densification behavior of the powders at different milling time. The hardness of the dense sintered structures was measured using a Vickers hardness tester. To determine mechanical strength under compressive loading, the sintered specimens were tested in a fully automated tensile tester from Instron (Model 3369) with a constant crosshead speed of 1mm/min.

RESULTS AND DISCUSSIONS
Phase Analyses

Figure 1 shows the XRD patterns for the powder mixture of $Ca(OH)_2$ and P_2O_5 that was subjected to mechanical activation for various time periods (in the range of 2 min to 3 h), together with the XRD pattern of the initial powder mixture, which was not subjected to any mechanical activation. The unmilled as-mixed powder shows only $Ca(OH)_2$ phase and was confirmed using JCPDS file no 9-432. With continued milling, the X-ray diffraction peaks became broader and their peak intensities decreased. The peak broadening is attributable to crystallite (grain) size refinement and accumulation of the lattice strain in the powder during milling. This trend continued up to 30 min and most of the initial material phase exhibited broad peaks indicating formation of a nanocrystalline structure or amorphous phase.

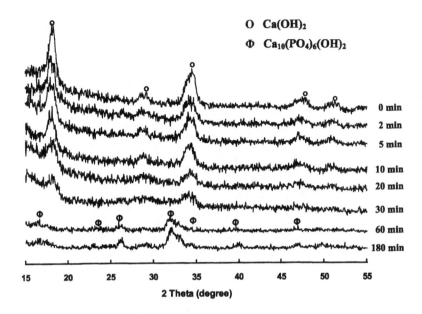

Figure 1. XRD patterns of the powder mixture of Ca(OH)$_2$ and P$_2$O$_5$ after milling for different times.

Low intensity peaks of the hydroxyapatite Ca$_{10}$(PO$_4$)$_6$(OH)$_2$ phase were observed even after milling for 1 h. No elemental Ca(OH)$_2$ peaks were seen at this stage, suggesting that mechanochemical solid-state reaction proceeded when the interface between the two solids came into close intimate contact and sheared by the mechanical force. After milling for 3 h, the diffraction pattern showed the same Ca$_{10}$(PO$_4$)$_6$(OH)$_2$ phase but with increased peak intensity. However no additional increase in intensity was noted after 3 h of milling, suggesting that the synthesis reaction was complete between 1 and 3 h. From this it could be concluded that 1 h of milling was sufficient for the homogeneous formation of the Ca$_{10}$(PO$_4$)$_6$(OH)$_2$ phase. In comparison to the earlier investigations[13,14] formation of the Ca$_{10}$(PO$_4$)$_6$(OH)$_2$ phase took place in a shorter time and without the formation of any transitional intermediate phases. This result indicates that the reaction between Ca(OH)$_2$ and P$_2$O$_5$ progressed rapidly because of the high mechanical milling energy.

Microstructural Analysis

Figure 2 shows the typical morphology of the Ca$_{10}$(PO$_4$)$_6$(OH)$_2$ composition powders milled for 3 h, as observed by SEM. The agglomerated powders have a size range of approximately 1 μm, a smooth surface and spherical shape. In addition, it was observed that some of the very fine particles were agglomerated. To obtain better information regarding the size and shape of the powder particles, TEM was used. Results of TEM analyses are shown in Figure 3. It is apparent from Figure 3 that mechanical milling for 3 h resulted in a significant refinement in the powder particle size due to the severe stress during mechanical alloying. TEM micrographs revealed that the synthesized powders were in the size range of 20-60 nm.

Figure 2. Scanning electron micrograph of the $Ca_{10}(PO_4)_6(OH)_2$, HAp, powder after milling for 3h.

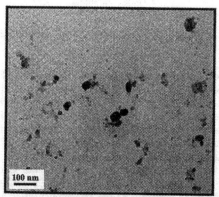

Figure 3. Transmission electron micrograph of the $Ca_{10}(PO_4)_6(OH)_2$ HAp, powder after milling for 3h.

Densification Studies

Green ceramic structures, prepared via uniaxial pressing using powder mixtures of $Ca(OH)_2$ and P_2O_5 after different milling times (0 min, 30min, 60 min and 180 min), were subjected to pressure-less sintering in a muffle furnace at 1300 °C for 6 h in ambient atmosphere for densification studies. Sintered ceramic structures were measured for their bulk-sintered density and the average sintered density for different milling times (0 min, 30min, 60 min and 180 min) was calculated and plotted as a function of milling time in minutes (Figure 4). It is

36

evident from this figure that the highest sintered density of 3.17 g/cc was recorded for powders milled for 180 min, corresponding to crystalline nano-grained hydroxyapatite.

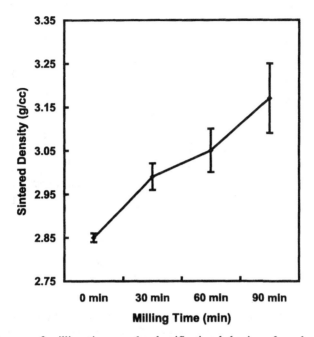

Figure 4. Influence of milling time on the densification behavior of mechanically alloyed $Ca(OH)_2$-P_2O_5 powders. Sintering was done at 1300 °C for 6 h.

Hardness Testing

A Vickers hardness tester was used to determine the hardness of sintered ceramic structures processed via uniaxial compaction using $Ca(OH)_2$-P_2O_5 powder mixture at different milling time (0 min, 30 min, 60 min and 180 min). The specimens were sintered at 1300 °C for 6 h. Results of hardness testing are presented in Figure 5 as a function of the sintered density for each of these milling times.

It can be observed from Figure 5 that an increase in the milling time has significantly improved the hardness of the $Ca(OH)_2$-P_2O_5 powder mixture. The highest hardness value was obtained for powders milled for 180 min corresponding to the formation of the $Ca_{10}(PO_4)_6(OH)_2$, HAp, phase. It is evident from the figure that there is a considerable rise, approximately 50 %, in hardness value for the nano HAp powders obtained after 180 min of milling compared to the initial composition, $Ca(OH)_2$-P_2O_5 powder mixture at 0 min. Figure 5 also reveals that hardness increased with the increase in density.

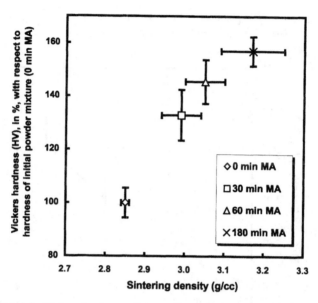

Figure 5. Variation in Vickers hardness of sintered compacts of $Ca(OH)_2$-P_2O_5 mixture after 0 min, 30 min, 90 min, and 180 min of mechanical alloying.

Compression Testing

    Uniaxial compression testing was conducted to evaluate and compare the mechanical properties of the ceramic powder mixture after 0 min, 30 min, 60 min and 180 min of milling. Cylindrical samples with the diameter to height ratio of 1:1.5 were prepared via uniaxial pressing in a steel mold followed by sintering at 1300 °C for 6 h. A set of 4 samples for each time point was tested. Results from compression testing are shown in Table 1. The highest compressive strength of 85.4±4.2 MPa was recorded for nano HAp ceramic structures (180 min of milling)

Table 1. Compressive strength of the sintered ceramic structures.

Type	Compression Strength (MPa)
$Ca(OH)_2$-P_2O_5 , 0 min	55.1 (±3.8)
$Ca(OH)_2$-P_2O_5, 30 min	61.7 (±4.6)
$Ca(OH)_2$-P_2O_5, 60 min	74.7 (±3.4)
$Ca(OH)_2$-P_2O_5, 180 min	85.4 (±4.2)

CONCLUSIONS

Nanocrystalline hydroxyapatite $(Ca_{10}(PO_4)_6(OH)_2$, HAp) ceramic powders were synthesized via mechanical solid-state reaction of a stoichiometric mixture of calcium hydroxide $(Ca(OH)_2)$ and phosphorous pentoxide (P_2O_5). A homogenous single phase of hydroxyapatite could be obtained in the powder milled for 180 min. The hydroxyl group needed for the formation of hydroxyapatite during the heat-treatment was supplied from the mechanochemical reaction between the starting powders during high-energy milling. TEM results showed that as-synthesized HAp powders were in the range of 20-60 nm. The MA process is very simple and economical which makes it highly suitable for mass production of nanocrystalline hydroxyapatite.

REFERENCES

[1] D.M. Liu, Q. Yang, T. Troczynski and W.J. Tseng, "Structural Evolution of Sol-Gel Hydroxyapatite", *Biomaterials*, 23, 1679-1687 (2002).

[2] A. Jillavekatasa and R.A. Condrate Sr., "Sol-gel Processing of Hydroxyapatite", *J Mater. Sci.* 33, 4111-19 (1998).

[3] A.T. Kuriakose, S. Narayana and B. Kalkuraa., "Synthesis of Stoichiometric Nano crystalline Hydroxyapatite by Ethanol-Based Sol–gel Technique at Low Temperature", (2004).

[4] P. Shuk, W. L. Suchanek, T. Hao, E. Gulliver, and R.E. Riman, "Mechanochemical-Hydrothermal Preparation of Crystalline Hydroxyapatite Powders at Room Temperature." *J. Mater. Res.*, 16, 1231-34 (2001).

[5] B. Yeong, X. Junmin, and J. Wang, "Mechanochemical Synthesis of Hydroxyapatite from Calcium Oxide and Brushite, *J. Am. Ceram. Soc.*, 84, 465-67 (2001).

[6] W. Kim, Q. Zhang and F. Saito, *J. Mater. Sci.*, 35, 5401-5405 (2000)

[7] J.S. Benjamin, "Dispersion strengthened superalloys by mechanical alloying", *Metall. Trans.* 1, 2943-51 (1970).

[8] C. Suryanarayana, "Bibliography on Mechanical Alloying and Milling", *Cambridge International Science Publishing*, Cambridge, UK, (1995).

[9] C. Suryanarayana, "Mechanical alloying and milling", *Prog. Mater. Sci*, 46, 1-184 (2001).

[10] C. Suryanarayana, "Mechanical Alloying and Milling", *Marcel-Dekker*, New York, NY (2004).

[11] J. Aoki, H. Akao and K. Kato, "Mechanical properties of sintered hydroxyapatite for prosthetic applications.", *J. Mater. Sci.*, 16, 809 (1981).

[12] J.G.C. Peelen, B.V. Rejda and K. de Groot, "Preparation and properties of sintered hydroxyapatite", *Ceramurgia Int*, 4, 71 (1980).

DRY HIGH SPEED MILLING AS A NEW MACHINING TECHNOLOGY OF CERAMICS FOR BIOMEDICAL AND OTHER APPLICATIONS

Prof. Prof. h. c. Dr. Anthimos Georgiadis
University of Lueneburg
Automation Technology
Volgershall 1
Lueneburg, Germany, 21339

Dr. Elena Sergeev
University of Lueneburg
Automation Technology
Volgershall 1
Lueneburg, Germany, 21339

ABSTRACT

Ceramics are the first choice for orthopaedic, dental and other biomedical applications. Because of the processes used, the conventional production of ceramic parts is moreover limited to materials that are not optimal from a medical point of view. Further critical issues for biomedical applications are cooling lubricants and polishing materials used for machining the ceramics. Absorption into the surface of the machined ceramic parts makes them not neutral for the human body so they can cause allergic reactions or diseases. The developed dry, high speed milling of ceramics (Ceramill) presented in this work solves this problem. The method and a prototype machine tool for 3D treatment with the implementation of advanced controls, CAD-CAM coupling, high speed spindle, the dry process and the dedicated new tools have been developed and approved in field tests. Work pieces with dimensions till 200x400x400 mm has been constructed. Suitable geometries and coatings for new milling tools have been determined. The machining tests were performed with various parameters such as diamond coating thickness and composition. The trials have been performed with aluminium oxide, zirconium oxide and silicon nitride. Ceramill leads to accurately finished products with tolerances of the order of magnitude of 1 micron and surface roughness of 0.2 microns. First measurements of the Weibull factor for silicon nitride show an encouraging value of 57 by a load of 628 MPa. A first model for the process based on the mechanistic approach has been established on the MatLab platform.

INTRODUCTION

Ceramics have been increasingly used in automotive, aerospace, military, biomedical and other applications. Despite the existence of many machining methods for ceramics, high costs and subsurface damage or other influences of the material still impede the use of ceramic products. The main objective of this paper is the presentation of a new approach for treatment of ceramics and other brittle, hard materials, which looks like the known dry, high speed milling of metals. (We call it Ceramill). The new treatment helps to overcome many of the appearing problems.

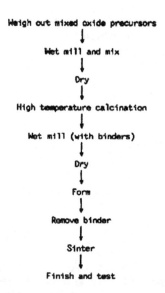

Weigh out mixed oxide precursors

↓

Wet mill and mix

↓

Dry

↓

High temperature calcination

↓

Wet mill (with binders)

↓

Dry

↓

Form

↓

Remove binder

↓

Sinter

↓

Finish and test

Figure 1: Standard mixed oxide processing route

The conventional ceramic processing technology consists typically of several steps [10] shown in Fig. 1. Furthermore, the conventional manufacturing of dentures is based upon single-piece production involving various cost-intensive manual stages. Ceramic processing proceeds stepwise, with each new step acting upon the results of the preceding step. If reproducibility and reliability of the end product are to be realized, it is clear that:

a) The process should contain as few steps as possible.
b) The process should contain checks at each step in order to monitor the process and ensure that the product properties are within acceptable limits.
c) Most careful control is needed during the earliest stage of the process as the cumulative effect of variations here is of great significance on final production properties.[10]

Ceramill abridges the conventional ceramic machining chain for the following chain links:

1. Wet mill with lubricants
2. Drying
3. Remove binding
4. Finishing and grinding after the sintering
5. Cleaning

The dry ceramic machining treatment has a specific importance for the medical applications. The lubricant residues on the surface of part that was not completely removed during the manufacturing process cause the unwanted allergic reactions in the human body. The cleaning of ceramics parts for medical implantations, after machining with lubricants, needs sometimes more costs and time as the machining itself. During Ceramill no lubricants or other materials can be absorbed because it is a dry process. Furthermore, there is no measurable increase of the temperature appears, which could induce degradation of the ceramic. In this paper, the process the prototype machine tools, the new tools and first results of finished parts will be decrypted, applied on aluminium, zirconium oxides and silicon nitride. An advanced high speed spindle and a new process monitoring system have been also investigated and implemented to the machine tool. These components will be shortly presented in the paper.

Modeling the Ceramill process.

A number of different methods to predict cutting forces have been developed over the last years for milling metals. These models can be classified into three major categories: analytical, empirical, and mechanistic methods. All of them are based on the plastic deformation and chip thickness analyses.

1. Analytical approaches [1,2] - model the physical mechanisms that occur during cutting. This includes complex mechanisms such as high strain rates, combined elastic and plastic deformations and it's not yet completely solved even for metal.
2. Empirical methods, where a number of machining experiments are performed and performance measures such as cutting forces, tool life, and tool wear are measured and regression model have to be built [3,4]. This method could be applied for high speed modeling of brittle materials, but it is connected with large experiments effort.
3. Mechanistic models, [1,5,6,7], predict the cutting forces based on a method that assumed cutting force to be proportional to the chip cross-section area.

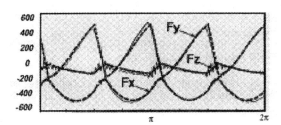

Figure 2. Simulated and measured cutting forces in end milling of metal. Milling cutter with 4 flutes [1]

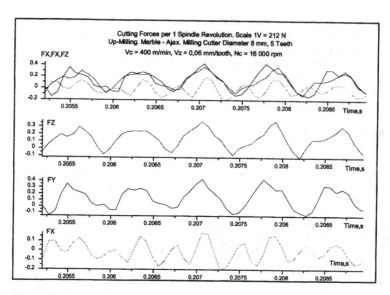

Figure 3. Cutting forces per one spindle revolution measured by marble milling

Although the cutting forces using ceramill on brittle materials(Fig. 3) are similar to those on metals (Fig. 2), the differences between them don't allow the direct application of models for milling of metals in the case of brittle materials like ceramic. In our case dynamic chip thickness and combined elastic and plastic deformations used in force calculations in well-known machining process simulation software (CutPro, Metalmax) is not occurred. It is evidently, there are completely different analytical approaches in the brittle materials milling. An approach based on the mechanistic process model have been used in order to develop a model predicting cutting forces and spindle true power by different cutting conditions.

According the mechanistic metals milling model [1] tangential $F_t(\varphi)$, radial $F_r(\varphi)$ and axial $F_a(\varphi)$ cutting forces are expressed as functions of varying uncut chip area $ah(\varphi)$ and depth-of-cut a:

$$F_t(\varphi) = K_{tc}ah(\varphi) + K_{te}a,$$
$$F_r(\varphi) = K_{rc}ah(\varphi) + K_{re}a,$$
$$F_a(\varphi) = K_{ac}ah(\varphi) + K_{ae}a,$$

where K_{tc}, K_{rc}, K_{ac} are the force coefficients contributed by the shearing action in tangential, radial, and axial directions and K_{te}, K_{re}, K_{ae} are the edge constants. K_{tc}, K_{rc}, K_{ac} coefficients correspond the shearing force itself, and K_{te}, K_{re}, K_{ae} characterise the ploughing at the flank of the cutting edge.

The first approach of the dry high speed milling mechanistic model for the brittle materials has been realized in MatLab. At the Figure 4. are shown the measured and predicted, calculated with process model cutting forces. The constant middle value of specific cutting forces per one spindle revolution has been taken into account.

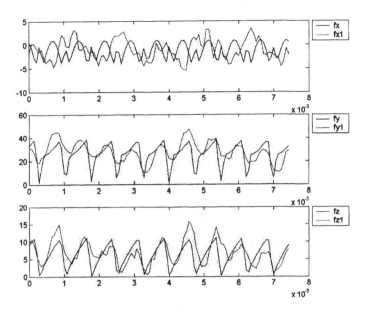

Figure 4.Predicted FX, FY, FZ and measured FX1, FY1, FZ1 cutting forces

The comparison between experimental and semi empirically calculated forces show a good agreement between them in the directions X, Y, Z. However, the model has to be improved in order to deliver reliable values for each combination of processing and material parameters .The following additional steps should be undertaken for a better understanding of the process:
1.The influence of the cutting parameters on the cutting forces must be taken into account. The specific cutting forces decreases when cutting speed increases, and specific cutting forces increases when feed rate per teeth increases.

2. Cutter deflection and run out has to be also taken into account. These parameters are not involved in presented model, and it is one of the possible reasons of notable differences between the measured and the predicted cutting forces.

3. Specific properties ceramics have to be investigated and implemented into the model. The presented state of the art in the area of dry high speed milling modeling shows that the constant specific cutting forces are not enough to characterize the material effect on the cutting process.

Unfortunately these additional steps, which significantly extend the process modelling task could not be done within the frame of this work. We plan to continue this research in the next project. Since January 2003 we are able to use the machine for tests. Nevertheless, our trials of milling ceramics are very encouraging as show bellow.

Experimental set up

Figure 5. The machine tool (LithoCerm1) for dry, high speed milling of ceramic (Ceramill)

A 5 axis machine tool (LithoCerm1) has been constructed dedicated for dry, high speed treatment of brittle materials (maintenance free [15]) with the following specifications:

Stroke x axis: 400 mm, max. acceleration: 1 g, max. speed: 15 m/min
Stroke y axis: 600 mm, max. acceleration: 1 g, max. speed: 15 m/min
Stroke z axis: 200 mm, max. acceleration: 1 g, max. speed: 15 m/min
C axis: 360°, max. speed 60 min-1
A axis: 0...90°, max. speed 60 min-1
Cutting force: max. 1.500 N
Tool diameter: up to 10 mm
Tool length: up to 50 mm
Work piece weight: 5 kg
Integration of a new high speed spindle, [11]
Adaptation and optimization of a 5 axis controller [13]

A general post processing system has been adapted to the 5-axis milling machine and linked to controller [13]. The postprocessor is capable to deliver tool orientation in joint angles, approach vectors or rotation angles of the world coordinate system. During the project a monitoring and guidance system was developed end tested. The system based on a PC board which can be integrated in the numeric control of the machine. Different sensors were integrated in the working space of the machine and a couple of tests were carried out. The different components of the used system are shown in the next table. Process and tool monitoring [14] was used possible with a controller integrated supervisor system. The system can easily integrated in the open control system of the applied PLC. The sensors could be true power or digital torque. Vibration can be used for tool wear and spindle monitoring. Another good quantity is the cutting force, which was manly used in our experiments but the integration of force sensors is very complicated and expensive for normal production plants. Another disadvantage is the short measuring time because of the sensor drift. Cutting of ceramics leads to very low process power, so here acoustic emission and vibration can be used.

Milling tools for ceramics

First the geometry of the tool has been investigated using tools from [12] like shown in Tab.1 In the next step nine combinations of diamond coating and basic hard metal (carbide grade) have been developed [12]in order to optimize milling tools for ceramics.

Type	∅	Rake Angle	Relief angle First	Second	Corner Radius	Z	Helix
JH910	6	12	14	25	0,3	3	40
JH930	6	4	13	18	0,2	5	41
JH930	6	4	13	18	0,5	5	41
JH120	6	2	14	25	0,6	4	20
JH130	6	-25	8	13	0,2	5	44

Table 1: Milling tools used for the optimization of the geometry

Pre-tests with ceramic B600 - Al2O3, Vickers hardness 1600, have shown that the following machining parameter would be useful for milling tools endurance tests:

Cutting Speed: 50 m/min
Feed Rate per Tooth: 0,06 mm
Depth-of-cut ap: 2 mm
Width-of-cut ae: 0,02 mm

A half pipe machining strategy turn out to be beneficial for the tool since not only the tool tip but also whole cutting edge works during the process. The geometry of JH930, Tab. 1, has shown the best results working on marble. Keeping this geometry 9 different coatings have been tested in order to increase the life time of the tools and improve the results. The coating is a crystalline diamond coating with a grain size in the nanometer range (20-100 nm). The characteristic of this coating is the low surface roughness through the extremely small grain size in the range of 20-100 nm. Through the high content of grain boundaries in relation to volume this coating is extremely ductile with the extreme high hardness of diamond.
The summary of ceramics milling endurance tests is shown below in the Table 2:

Ceramic	Vickers hardness	Stability Cutting Time, min	Stability Cutting Distance, m
Aluminium oxide	1600	55,35	46,56
Zircon oxide	1250	251,7	117,78
Silicon Nitride	1650	5,58	4,02

Table 2 Results of ceramics milling endurance tests

 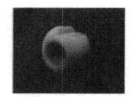 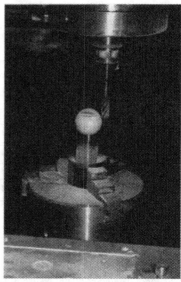

Figure 6. Ceramill treatment on aluminum oxide implantations. Left hand pieces after cutting and right hand during the process.

Ceramics parts

The first trials realized at the LithoCerm1 have shown that high speed milling on ceramics is feasible (Figure 6) The material of the work piece (hip joint) was Al$_2$O$_3$ with Vickers hardness 1800. A milling tool total stability time of about 10 min has been achieved.
Ceramics milling endurance tests have been performed with the parts shown below: a) aluminum oxide B600, HV 1650 (Figure 6 and 7), b) zircon oxide (Figure 8), and c) silicon nitride (Figure 9).
The surface quality of machined ceramic slabs is: Rz = 6,3 – 8,0 m,
Ra = 0,95 - 1,04 m; corresponds milling mode finished, superfine.
The maximal cutting forces by the B600 ceramics milling reaches 150 to 200 N for the new milling cutters and 300 – 450 N for the worn out milling cutter at the end of the test. The middle cutting forces for the 4mm thick ceramics slabs are about 20% higher, as for the 10 mm thick marble slabs.

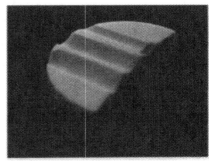

Figure 7 Ceramic (Al2O3) milling at LithoCerm (left); ceramic part after the machining (right)

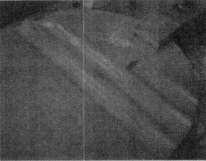

Figure 8. Zircon oxide milling at the LithoCerm1. Left – machining process, right machined part

The surface quality of zircon oxide slabs is: $Rz = 6,5 - 9,6$ µm, $Ra = 1,0 - 1,2$ µm; corresponds milling mode finished, superfine. Maximal cutting forces by the zircon oxide milling: $250 - 300$ N.

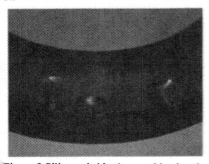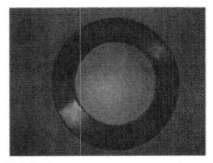

Figure 9 Silicon nitride ring machined at the LithoCerm1

The surface quality in this case is: Rz = 4,6 - 5 μm, Ra = 0,65 – 0,8 μm; corresponds milling mode finished, superfine. First tests for the influence of the ceramill on the ceramic have shown fantastic results of a Weibull factor of 57 by the load of 628 MPa.

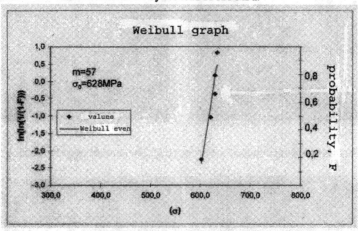

Figure.10 The Weibull graph for silicon nitride

CONCLUSIONS

The new machining process for hard, brittle materials (here called ceramill) promises to enable the production of biomedical implants with improved specifications. Using very advanced machine tools, diamond cutters and methods, till now only known in the metal treatment, high quality implants of ceramics and other products are possible in mass production. The lack of lubricants allows absolute clean parts. The very high Weibull factor is an indicator of a process without influences on the material. If further investigations will lead to similar factors Ceramill may reshape the whole sector of manufacturing ceramics and especially implants by allowing new dimensions of parts, free forms, new hard, brittle materials, more complex parts. The generation of CAD drawings out of images has been probed and it opens the production of individual prosthesis In order to reach the maximal achievements the model of the process has to be improved. The next model has to integrate the material parameters and the tool specifications and it has to optimize them. High speed video captures could help to understand the process. Determination of the sizes of the produced dust in combination of the surface and material specification will also help to calculate the energy flow during the process. Optimizing those parameters a complete model could help to reach stable production parameters for milling of ceramics.

ACKNOWLEDGEMENTS

We would like to thank all organizations, companies and people that have contributed to the LITHO-PRO (www.lithopro.org) project or supported it in any way. Especially we would like to thank the European Commission for funding the project. Prof. Dr. Gerold Schneider, TU HH, many thanks for the measurement of the Weibull factor.

REFERENCES

[1] Altintas. U.; "Manufacturing Automation", *Cambridge University Press, 1st edition*, (2000)
[2] Moufki, A., Dudzinski, D., Molinari, A. and Rausch, M., " Thermoviscoplastic modeling of oblique cutting: forces and chip flow predictions", *Int. J. Mech. Sci., Vol.42, pp. 1205-1232* (2000)
[3] Po-Tsang Huang, Joseph C. Chen, Dr. Chai-Yu Chou A " Statistical Approach in Detecting Tool Breakage in End Milling Operations" *Journal of Industrial Technology* • Volume 15, Number 3 (May 1999 to July 1999)
[4] Armarego, E. & Brown, J., " The Machining of Metals", Prentice-*Hall, New Jersey* (1969).
[5] Kline,W., DeVor, R. and Lindberg, J., " The Prediction of Cutting Forces in End Milling", *Int. J. Mach. Tool. Des. Res.*, Vol.22/1, pp. 7-22 (1982).
[6] Altintas, U. & Spence, A.," End Milling force algorithm for CAD Systems", *Annals of CIRP*, Vol.40/1, pp. 31-34 (1991).
[7] Jayaram, S., Kapoor, S. and DeVor, R., " Estimation of the Specific Cutting Pressures for Mechanistic Cutting Force Models", *Int. J. Mach. Tool. Man.*, Vol.41, pp. 265-281 (2000)
[8] König, W and Essel, K " Spezifische Schnittkraftwerte für die Zerspanung metallischer Werkstoffe". *VDE Verlag Stahleisen mbG*, Düsseldorf (1973)
[9] Eberhard Paucksch „Zerspantechnik", *Vieweg Verlag*, Braunschweig (1992)
[10] Advanced Ceramic Processing and Technology Edited by: Binner, J.G.P. © 1990 *William Andrew Publishing/Noyes* (1990)
[11] IBAG-Spindle HF100 AI 50 C, IBAG GmbH
[12] Jabro-Tools GmbH
[13] Siemens 840Di
[14] ARTIS-CTM, ARTIS GmbH
[15] SNR GmbH

Biomaterials, Performance and Testing

NANOCERAMICS INTERCALATED WITH Gd-DTPA FOR POTENTIAL IMAGING OF SYSTEMS *IN VIVO*

Seo-Young Kwak and Waltraud M. Kriven
Department of Materials Science and Engineering
University of Illinois at Urbana-Champaign,
Urbana, IL 61801, USA

Robert B Clarkson[†]
Department of Veterinary Clinical Medicine,
University of Illinois at Urbana-Champaign,
Urbana, IL 61801, USA

Benjamin J. Tucker and R. Linn Belford
Department of Chemistry, Illinois Electron Paramagnetic Resonance Center,
University of Illinois at Urbana-Champaign,
Urbana, IL 61801, USA

ABSTRACT

Nano–sized ceramic particles, such as layered double hydroxides (LDHs) have been developed to preserve and deliver drugs or genes in the body. In this report, we describe a new ceramic contrast reagent for magnetic resonance imaging (MRI). It has been developed by intercalation of the chelate diethylenetriamine-N, N, N', N", N"'-pentaacetate (DTPA) complexed with Gd^{3+} (Gd-DTPA). Its characteristic electron paramagnetic resonance (EPR) spectrum shows a slow tumbling, space-limited environment in the LDH matrix. Gd-DTPA LDH is designed to track the route of an intercalated drug and to be applied as a new nanoceramic contrast reagent for MRI in order to detect any image tumors and their dendrites. According to XRD, SEM and TEM, the nanohybrid crystal structure and morphology possess typical, two-dimensional metal ion layering. Elemental analysis (CHN and ICP) shows a stable Gd-DTPA complex in the interlayers of LDH. This idea has the potential to extend the use of nanoceramics for gene or drug delivery.

INTRODUCTION

Recently, biomolecules such as nucleotide, deoxyribonucleic acid, pentose, amino acid, polypeptide and anticancer drugs have been intercalated into LDHs to investigate their use as a novel, tunable, drug delivery system.[1-8] This has created interest in an approach involving inorganic solid particles and an interlayer contrast agent for monitoring *in vivo* systems. Applying intercalated LDH as a therapeutic drug delivery carrier requires appropriate *in vitro* and *in vivo* research. In this paper, we describe a new potential ceramic transport mechanism of contrast agents in MRI and a method to study drug delivery via LDH particles intercalated with Gd-DTPA complexes. Gd^{2+}-DTPA was approved by the FDA in 1988; over 30 metric tons of gadolinium have been administered to millions of patients worldwide.[9] Currently, approximately 30 % of

---

[†] Deceased

---

magnetic resonance imaging (MRI) exams include the use of contrast agents; this is projected to increase as new agents and applications arise.[10] Gd-DTPA is used widely in clinical MRI (magnetic resonance imaging) for humans and animals. Clinicians desire enhanced stability, a neutral molecule, and lower osmolality, viscosity, and chemotoxicity in contrast agents.[10] Gd-DTPA is generally safe, stable, and necessary only in low concentrations (1 mM). However, it is low in effective charge (1 mol e^-/226 g), and thus impossible to deliver through non-polar cell membranes.

In this study, the LDH used is $[Mg_3Al(OH)_8]OH \cdot nH_2O$. Some reported methods for the intercalation of species with this LDH are an ion-exchange reaction,[11-13] a co-precipitation method,[14] a calcination-rehydration reaction,[15-19] etc. The synthesis of this intercalate has not been straightforward, because the LDH layers are basic, and the Gd-DTPA complex has a bulky molecular structure and low charge to mass ratio. The molecular dimensions have been calculated using covalent bond lengths of C-C (1.50-1.55 Å), C=C (1.34 Å), C-H (1.08-1.10 Å), C-N (1.47 Å), C=N (1.27 Å), C=O (1.21 Å), N-H (1.00 Å) and O-H (0.96 Å), etc. Fig. 1 is a schematic of the molecular structure of Gd-DTPA.[9]

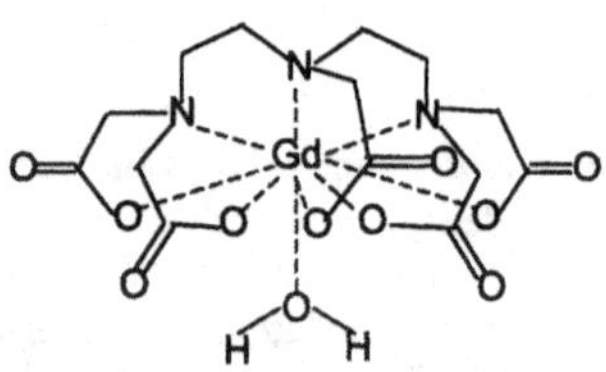

Fig. 1. The schematic molecular structure of Gd-DTPA.[9]

EXPERIMENTAL PROCEDURES

We prepared the LDHs using similar methods to those described by Yun and Pinnavaia.[16] An aqueous solution of 0.3 M $MgCl_2$ and 0.1 M $AlCl_3$ was titrated dropwise with a 0.1 M NaOH solution; the resulting solution was aged with vigorous stirring in air for 48 h. This resulted in a pristine, carbonated LDH, due to carbon dioxide from the air intercalating between the metal ion layers. After washing and drying, the carbonated LDH was calcined at 773K for 5 h. The oxide was dispersed in distilled, deionized water (1.0 wt %) and stirred at room temperature for 5 days, to form the rehydrated LDH. The LDH suspension was added to an adipic acid solution in 2-fold excess of its calculated anionic exchange capacity (AEC) to form the LDH adipate. After stirring for 1 h at 50°C, the suspension was allowed to settle and then decanted, leaving a slurry of the LDH adipate. 100 mL of degassed boiling water was added to the slurry. The hot LDH slurry was then added slowly to Na_2Gd-DTPA solution in 3-fold excess of the AEC at 100 °C in a sealed flask. H_2Gd-DTPA was previously titrated with NaOH solution above pH 4 to produce the Na_2Gd-DTPA solution.[9] The reaction mixture was stirred for an additional 72 h at the reflux temperature. The resulting product was washed with water and dried in air.

RESULTS

The XRD patterns of the carbonated, calcined, rehydrated and dried Gd-DTPA LDH species are shown in Fig. 2. The main diffraction peak ($d_{003} = 0.76$ nm) corresponds to Fig. 2(a), which was clearly the carbonated LDH. The disappearance of the d_{003} diffraction peak (Fig. 2(b)) shows that calcination destroyed the LDH structure. In Fig. 2, sharp diffraction peaks and expanded basal spacings of $d_{003} = 0.79$ nm (c) and $d_{003} = 1.46$ nm (d) proved that the rehydration and intercalation reactions occurred. Taking the thickness of a LDH layer to be 0.48 nm, the gallery heights were calculated to be 0.98 nm. This is consistent with the van der Waals diameter, estimated from crystallographic data for the Gd-DTPA complex, and drawn schematically in Fig. 1.[9] This value supports the expectation that the molecular plane of the Gd-DTPA complex was horizontally oriented in the hydroxide basal layer.

We found that the Gd-DTPA complex was not fully intercalated compared to the theoretical AEC of the pristine LDH. ICP (inductively coupled plasma) elemental analysis revealed the formula for the resulting compound to be $Mg_{0.71}Al_{0.29}(OH)_2^-$ $(Gd-DTPA^{2-})_{0.005}(OH^-)_{0.28} \cdot n\ H_2O$. The intercalation reaction depends on the relationship between the charge and molecular dimensions of the intercalant and the equivalent area of LDH. When Bravo-Suárez et al.[19] proposed an equation to estimate the average interpillar distance (IPD) among the LDH interlayer anion pillars, assuming total anion exchange, they used the unit cell parameters of the LDH, the charge of the anion and average diameter of the top view area of the intercalated anion. Gd-DTPA has a large top view area (~ 140 Å^2) and the charge is only 2^-. The unit cell area of LDH is 8.32 Å^2, so the equivalent area is 28.3 Å^2/e$^-$. Approximately one Gd-DTPA will occupy 5 equivalent areas of LDH. In order to compensate for the excess charge of LDH, OH$^-$ ions are expected to remain in the LDH interlayer. Due to this low effective charge of the Gd-DTPA complex, the anion behaves more like a molecule than like ions with higher charge to mass ratio. Thus, this calcination-rehydration method is one way to synthesize the hybrid.[19]

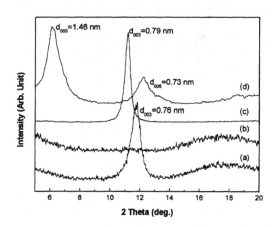

Fig. 2. XRD patterns (Cu-Kα) of the carbonated LDH (a), calcined LDH (b), rehydrated
LDH (c), and Gd-DTPA LDH (d)

Fig. 3 SEM micrograph of the Gd-DTPA-LDH

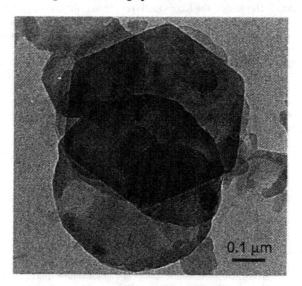

Fig. 4 TEM micrograph of the Gd-DTPA-LDH

The SEM micrograph of Gd-DTPA LDH in Fig. 3 shows well-dispersed,
hexagonal platelets, resembling the morphology of typical LDHs.[20] These platelets had
an average thickness of 19 ± 4 nm and a diameter of 121 ± 19 nm.   Since the initial
morphology was retained up to the final product, the synthesis route is likely to proceed
by a topotactic reaction.   TEM micrographs (Fig. 4) of the sample revealed the two-

dimensional layered structure of this material. Well-formed hexagonal or rounded platelets were observed in agreement with SEM findings. Large particles (approximately 550 nm in diameter) were selected in the TEM to clearly show the discrete shapes, although the majority of particles were ~121 nm. For SEM, the Gd-DTPA LDH suspension was dropped on to a Si-wafer and dried at room temperature.

The Gd-DTPA LDH particles gave a strong, clear, typical EPR signal similar to Gd-DTPA signals in aqueous solution at room temperature (Figs. 5 and 6). The spectra indicate that Gd-DTPA is much more concentrated in the LDH than would be necessary for MRI detection. However, there was also a larger concentration in the solution around the LDH particles. Unfortunately, the LDH particles leaked out some Gd-DTPA, as after a few washings, the Gd-DTPA signal was a mixture of bound Gd-DTPA in the LDH particles and free Gd-DTPA in solution. After several washings, both Gd-DTPA signals were virtually nonexistent. The X Band (Fig. 5) and W Band (Fig. 6) spectra below show Gd-DTPA in LDH signals after an optimal number of washings for each (meaning that the concentration of the samples inside and outside of the particles was different). This was determined by analyzing suspensions of the samples after stirring (by hand), centrifuging the samples in water, and by analyzing the supernatants. They X and W band spectra each compared subtracted spectra (bound) to supernatant (free) spectra.

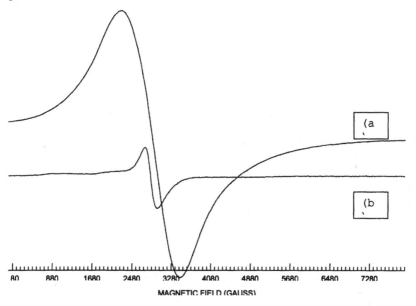

Fig. 5: X Band (9.5 GHz) EPR spectra of Gd-DTPA in LDH (b) and free Gd-DTPA (a)

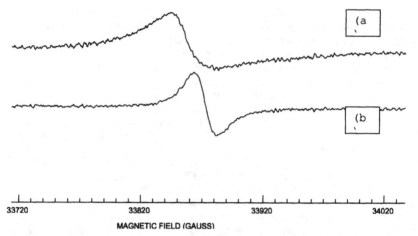

Fig. 6. W Band (95 GHz) EPR spectra of Gd-DTPA in LDH (a) and free Gd-DTPA (b)

There was a marked narrowing of the bound Gd-DTPA signal in the X band spectra (Fig. 5), which compared to widening of the bound Gd-DTPA signal in the W band spectra (Fig. 6). A difference between the free and bound signal in MRI should be distinguishable. A higher concentration of Gd-DTPA in solution could cause more spin-spin interactions of the Gd III centers, so the magnetic field saw the Gd III centers in different energetic states, resulting in broadening. The difference could also be due to rotational constraint of the Gd-DTPA inside the LDH particles and the speed of interaction of waters moving by the Gd III center, or a combination of the above effects. The microwave frequency at which Gd resonance occurs depends on the magnetic field, the electronic and chemical properties of the molecule or complex, as well as the orientation of the Gd complex. If the Gd-DTPA complexes rotate sufficiently rapidly (as they do in aqueous solution) the different molecular orientations are averaged out, and the resonance peak is narrow. In general, the higher the microwave frequency of the spectrometer, the faster the molecular rotation required to produce comparable narrowing.

Pulsed NMR measurements on the Gd-DTPA in LDH particles and Gd-DTPA in aqueous solution showed longer times measured for Gd-DTPA in LDH particles than for Gd-DTPA in water. When water molecules move into the hydration shell of the Gd-DTPA, the chemical environment changes, causing spin-lattice relaxation, and there could also be spin-spin relaxation by neighboring hydrogens. This relaxation is thus much quicker when there is no hindrance to movement of water molecules into, away from, and past the Gd-DTPA coordination structure. These physical constraints are consistent with the size of the particles and the time it takes to hydrate the particles when they are dry.

CONCLUSIONS

We have investigated a possible new delivery system for a Gd-DTPA chelated complex. Since the complex Gd-DTPA has a bulky molecular structure and its charge to mass ratio is relatively low, it was difficult to intercalate it into the $[Mg_3Al(OH)_8]OH \cdot nH_2O$ LDH. The intercalation occurred through an established but indirect method. XRD diffractometry and EPR spectra of Gd-DTPA LDH showed that the drugs in the interlayer space of LDH retained their chemical and biological integrity. The content of interlayer Gd-DTPA was lower than expected, but the Gd-DTPA LDH particles may still be concentrated enough to use in MRI imaging. The Gd-DTPA bound to the particles was differentiable by EPR from free Gd-DTPA, so that MRI imaging should be able to differentiate free from bound Gd-DTPA. Further research will continue on the use of LDH particles as Gd-DTPA carriers for potential applications *in vivo* monitoring of drug delivery and its effects.

ACKNOWLEDGEMENTS
The work was partially carried out in the Center for Microanalysis of Materials, University of Illinois, which is partially supported by the U.S. Department of Energy under grant DEFG02-91-ER45439.

REFERENCES
[1]J. H. Choy, S. Y. Kwak, J. S. Park, Y. J. Jeong and J. Portier, "Intercalative Nanohybrids of Nucleoside Monophosphates and DNA in Layered Metal Hydroxide", *J. Am. Chem. Soc.*, **121**, 1399-1400 (1999).
[2]J. H. Choy, S. Y. Kwak, Y. J. Jeong and J. S. Park, "Inorganic Layered Double Hydroxide as a Non-Viral Vector", *Angew. Chem. Int. Ed.*, **39**, 4042-4045 (2000).
[3]J. H. Choy, J. S. Park, S. Y. Kwak, Y. J. Jeong and Y. S. Han, "Layered Double Hydroxide as Gene Reservoir", *Mol. Cryst. Liq. Cryst.*, **341**, 425-429 (2000).
[4]J. H. Choy, S. Y. Kwak, J. S. Park and Y. J. Jeong, "Cellular Uptake Behavior of $[\gamma\text{-}^{32}P]$ Labeled ATP-LDH Nanohybrids", *J. Mater. Chem.*, **11**, 1671-1674 (2001).
[5]V. Ambrogi, G. Fardella, G. Grandolini and L. Perioli, "Intercalation Compounds of Hydrotalcite-like Anionic Clays with Anti-inflammatory Agents - I. Intercalation and in Vitro Release of Ibuprofen", *Int. J. Pharm.*, **220**, 23-32 (2001).
[6]A. I. Khan, L. Lei, A. J. Norquist and D. Óhare, "Intercalation and Controlled Release of Pharmaceutically Active Compounds from a Layered Double Hydroxide", *Chem. Commun.*, 2342-2343 (2001).
[7]S. H. Hwang, Y. S. Han, J. H. Choy, "Intercalation of Functional Organic Molecules with Pharmaceutical, Cosmeceutical and Nutraceutical Functions into Layered Double Hydroxides and Zinc Basic Salts", *B. Kor. Chem. Soc.*, **22**[9], 1019-1022 (2001).
[8]K. M. Tyner, S. R. Schiffman and E. P. Giannelis, "Nanohybrids as Delivery Vehicles for Camptothecin", *J. Control. Release*, **95**, 501-514 (2004).
[9]P. Caravan, J. J. Ellison, T. J. McMurry, and R. B. Lauffer, "Gadolinium(III) Chelates as MRI Contrast Agents: Structure, Dynamics, and Applications", *Chem. Rev.*, **99**, 2293-2352 (1999).
[10]T. J. Manning, W. Fiskus, M. Mitchell and L. Dees, "Determination of the Protonation Constants of Gadolinium(III) Diethyltriaminepentaacetic Acid by Solvent Extraction and ICP-AES", *Spectro. Lett.*, **323** [3], 463-467 (1999).

[11]T. Kwon, G. A. Tsigdinos and T. J. Pinnavaia, "Pillaring of Layered Double Hydroxides (LDHs) by Polyoxometalate Anions", *J. Am. Chem. Soc.*, **110**, 3653-3654 (1988).

[12]E. D. Dimotakis and T. J. Pinnavia, "New Route to Layered Double Hydroxides Intercalated by Organic-Anion-Precursors to Poyoxometalate-pillared Derivatives", *Inorg. Chem.*, **29**, 2393-2394 (1990).

[13]J. Wang, Y. Tian, R. C. Wang and A. Clearfield, "Pillaring of Layered Double Hydroxides with Polyoxometalates in Aqueous Solution without Use of Preswelling Agents", *Chem. Mater.*, **4**, 1276-1282 (1992).

[14]F. Kooli and W. Jones, "Direct |Synthesis of Polyoxovanadate-pillared Layered Double Hydroxides" *Inorg. Chem.*, **34**, 6237 (1995).

[15]E. Narita, P. D. Kaviratna and T. J. Pinnavaia, "Direct Synthesis of a Polymetalate-pillared Layered Double Hydroxide by Coprecipitation", *J. Chem. Soc.-Chem. Commun.*, 60-62 (1993).

[16]S. K. Yun and T. J. Pinnavaia, "Layered Double Hydroxides Intercalated by Polyoxometalate Anions with Keggin(α-$H_2W_{12}O_{40}^{6-}$), Dawson(α-$P_2W_{18}O_{62}^{6-}$), and Finke($Co_4(H_2O)_2(PW_9O_{34})_2^{10-}$) Structures", *Inorg. Chem.*, **35**, 6853-6860 (1996).

[17]S. Aisawa, H. Hirahara, K. Ishiyama, W. Ogasawara, Y. Umetsu and E. Narita, "Sugar-anionic Clay Composite Materials: Intercalation of Pentoses in Layered Double Hydroxide", *J. Solid State Chem.*, **174**, 342-348 (2003).

[18]S. Aisawa, H. Hirahara, S. Takahashi, Y. Umetsu and E. Narita, "Stereoselective Intercalation of Hexose for Layered Double Hydroxide by Calcination-Rehydration Reaction", *Chem. Lett.*, **33** [3], 306-307 (2004).

[19]J. J. Bravo-Suárez, E. A. Páez-Mozo and S. T. Oyama, "Intercalation of |Decamolybdodicobaltate(III) Anion in Layered Double Hydroxides", *Chem. Mater.*, **16**, 1214-1225, (2004).

[20]W. M. Kriven, S. Y. Kwak, M. A. Wallig, and J. H. Choy, "Bio-resorbable Nanoceramics for Gene and Drug Delivery", *MRS Bulletin*, **29** [1] 33-37 (2004).

NANOPHASE HYDROXYAPATITE COATINGS ON TITANIUM FOR IMPROVED OSTEOBLAST FUNCTIONS

Michiko Sato[1], Marisa A. Sambito[2], Arash Aslani[2], Nader M. Kalkhoran[2], Elliott B. Slamovich[1], and Thomas J. Webster[1,3]

[1]School of Materials Engineering and [3]Weldon School of Biomedical Engineering, Purdue University, West Lafayette, IN 47907, USA
[2]Spire Biomedical, Corp., Bedford, MA 01730, USA

ABSTRACT

In order to improve orthopedic implant performance, the objective of this in vitro study was to synthesize nanocrystalline hydroxyapatite (HA) powders and use such powders to coat titanium. HA was synthesized through a wet chemical process. The precipitated powders were either sintered in order to produce UltraCap (or traditionally-used microcrystalline size). HA powders or they were treated hydrothermally to produce nanocrystalline size HA powders. Some of the UltraCap and nanocrystalline HA powders were doped with yttrium (Y) because previous studies demonstrated that Y-doped HA in bulk improved osteoblast (or bone-forming cell) function over undoped HA. These powders were deposited onto titanium by a novel room-temperature process, called IonTite™, which provides the advantage of retaining properties of the starting HA powders. The original HA particles were characterized using XRD, ICP-AES, BET, a laser particle size analyzer, and SEM. The properties of the resulting HA-coatings were compared to properties of the original HA powders. The results showed that the chemical and physical properties of the original HA powders were retained when coated on titanium by IonTite™, as determined by XRD, SEM, and AFM analysis. In addition, results showed increased osteoblast adhesion on the nanocrystalline HA IonTite™ coatings compared to traditionally used plasma-sprayed HA coatings. Results also demonstrated greater amounts of calcium deposition by osteoblasts cultured on Y-doped nanocrystalline HA coatings compared to either UltraCap IonTite™ coatings or plasma-sprayed coatings. These results encourage further studies on nanophase IonTite™ HA coatings for orthopedic applications.

INTRODUCTION

Hydroxyapatite (HA) coated titanium is a promising material for orthopedic applications because it has been shown to promote bone regeneration when compared to non-coated titanium[1]. Plasma spray deposition is widely used to coat titanium. However, this technique has severe limitations because the high temperature and rapid cooling that are associated with this process yield a variety of chemical phases and lower HA crystallinity[2]. This is a problem as HA dissolution rates increase with decreasing crystallinity[3] and other calcium phosphorous compounds, such as β-tricalcium phosphate and tetracalcium phosphate, dissolve faster than HA[4]. Moreover, previous studies[5] on bulk HA have suggested that particle features in the nano-phase regime enhance osteoblast (bone-forming cell) functions. Furthermore, stimulated osteoblast functions were also observed on bulk HA doped with yttrium (Y) compared to undoped HA[6]. It is essential to optimize osteoblast functions to accelerate new bone formation between implants and juxtaposed bone. Moreover, nanophase HA is also found in natural bone, where nanoscale apatite-like crystals mineralize collagen fibers[7].

Due to the novelty of nanophase HA in increasing bone regeneration, it is of utmost importance to develop coating techniques that will deposit nano HA on a metal surface. Plasma spray deposition will not deposit nanometer crystals onto a metallic surface, rather it will produce amorphous calcium phosphate derivatives with micron crystal sizes. In this study, the first objective towards this goal was to synthesize nano-sized HA with and without doped Y ions and characterize the resulting HA particles that were to be coated on titanium. Moreover, HA-coated titanium samples were characterized and compared to the original HA particles. Such information is useful to elucidate changes in chemistry, crystallinity, and particle size that may have occurred during the coating process. Lastly, osteoblast functions were measured on titanium coated with the various HA particles.

EXPERIMETAL PROCEDURE

UltraCap (or microcrystalline) HA powders

An ammonium phosphate (Sigma) solution was added into the water at room temperature that had been adjusted to pH 10 using ammonium hydroxide. Then, a 1M calcium nitrate (Sigma) solution was added at a rate of 3.6 ml/min. Precipitation occurred as soon as the calcium nitrate was added, and continued for 24 h. The precipitates were centrifuged and rinsed with water twice, and subsequently, the precipitates were dried at 80 °C overnight in an oven. The dried precipitates were sintered at 1100 °C for 1h at a ramping rate of 1075 °C /h and were then crushed with a mortar and a pestle.

Y-doped UltraCap HA powders were prepared in a similar manner except that Y nitrate hexahydrate (Aldrich) of 5 wt. % relative to HA (2.8 mol% of Ca) was dissolved in the 1M calcium nitrate solution.

Nanocrystalline HA powders

HA powders were precipitated as described above, the precipitation continuing for 10 min at room temperature. Then some of the supernatant was removed by centrifuging one time to reduce the suspension volume by 75%. The concentrated HA-precipitated aqueous solution was added into a 125ml Teflon liner (Parr Instrument). The Teflon liner was sealed tightly in an autoclave (Parr Acid Digestion Bombs 4748; Parr Instrument) and processed hydrothermally at 200°C for 20 h. The hydrothermal method of HA synthesis is described in detail by Ioku et al[8]. High crystallization is achieved at relatively low temperatures but under a higher pressure than atmospheric. As a result, nano-sized HA crystallites can be obtained. After hydrothermal treatment, the HA particles were rinsed with dH$_2$O two times and dried at 80°C overnight in an oven.

Y-doped nanocrystalline HA powders were prepared in the same way described previously except for adding 1M calcium nitrate solution with dissolved Y nitrate hexahydrate of 5 wt. % relative to HA dropwise to the ammonium phosphate solution.

HA powder characterization

All powders were characterized by X-ray diffraction using Cu-Kα radiation (Siemens D500 Kristalloflex; Bruker AXS Inc.), a scanning electron microscope (SEM; JEOL JSM-35CF SEM), and inductively coupled plasma-atomic emission spectroscopy (ICP-AES) conducted by Analytical Impact to determine the Ca/P ratio of the HA synthesized. Particle size was determined by a laser particle size analyzer (COULTER LS 230; Coulter Corporation, FL). A BET surface area analyzer (SA3100; Beckman Coulter) was used to measure the surface area of the particles and crystallite size was determined using a theoretical HA density of 3.16 g/cm^3.

IonTite^{TM} coatings

The undoped and doped UltraCap and nanocrystalline HA powders were deposited on titanium substrates at Spire Biomedical, Inc. utilizing IonTite^{TM} processing.

For comparison, commercially available plasma-sprayed HA coatings (Himed) were deposited on the same titanium substrates used for IonTite^{TM} coatings.

The coatings were characterized with X-ray diffraction, SEM, and atomic force microscopy (AFM, Multimode SPM; Digital Instruments) to analyze surface morphologies, root mean square (RMS) roughness, and surface area. Scanning was conducted on five different places chosen randomly at a scanning rate of 1 Hz in tapping mode. Surface areas and RMS values were averaged and reported with standard deviation (SD).

Cell culture

Samples tested for cytocompatibility were: 1) IonTite^{TM} coated UltraCap HA, 2) IonTite^{TM} coated nanocrystalline HA, 3) IonTite^{TM} coated Y-doped UltraCap HA, 4) IonTite^{TM} coated Y-doped nanocrystalline HA, 5) plasma-sprayed HA, 6) uncoated titanium, and 7) glass. The glass substrates (Fisher Scientific, Inc.) etched with 1.0 N NaOH for 1h were prepared as a reference. 2,500 cell/cm^2 of bone forming cells, human osteoblast-like cells (ATCC; CRL-11372), were seeded onto samples in Dulbecco's Modified Eagle Medium (DMEM) containing 10% fetal bovine serum (FBS; Hyclone) and 1% penicillin/streptomycin (Hyclone). The cells were used without further characterization at passage numbers of 5-12. The samples were then incubated in an atmosphere of 5% CO$_2$/95% air at 37°C for 4 h. After the specified time period, non-adherent cells were removed by rinsing with phosphate buffered saline (PBS). The adherent cells were stained with calcein and ethidium homodimer (L-3224; Molecular Probes) then counted under a fluorescence microscope (Leica DMIRB; Leica Microsystems, Germany). Cell counts were completed at five random fields per cm^2. Experiments were run in triplicate and repeated three separate times.

Long-term osteoblast functions were also determined in this study. 50,000 osteoblasts/cm^2 of osteoblasts seeded on the samples were cultured for 21 days under the standard conditions described above to assess calcium deposition. DMEM (supplemented with 10% FBS, 1% penicillin/streptomycin, 50μg/ml L-ascorbate (Sigma), and 10mM β-glycerophosphate (Sigma)) was replaced every 2 or 3 days during culturing. DMEM for samples without cells was also replaced under the same condition. Total calcium (mg/cm^2) was calculated from standard curves of absorbance versus calcium concentration. Calcium concentration from

HA coatings without cells was subtracted from all values. Experiments were run in triplicate and were conducted during the same time period.

Osteoblast adhesion cell density and calcium concentration were represented by the mean value and standard error of the mean (SEM). T-tests were used to check statistical significance.

RESULTS AND DISCUSSION

Synthesized HA powders

The synthesized undoped and Y-doped UltraCap powders were composed of an HA phase (Fig. 1 (A) and (B)). Undoped and Y-doped nanocrystalline HA particles were also found to consist of a single HA phase (Fig. 1 (C) and (D)).

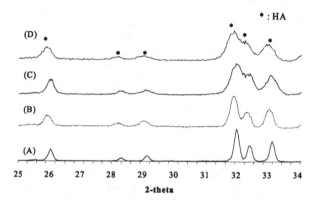

Fig. 1: XRD for (A) UltraCap HA powders, (B) Y-doped UltraCap HA powders, (C) nanocrystalline HA powders, and (D) Y-doped nanocrystalline HA powders.

The Ca/P ratio of UltraCap powders was 1.63 which was a little below the stoichiometric ratio of 1.67. The nanocrystalline HA powders exhibited a Ca/P ratio of 1.61. The Ca/P ratio of Y-doped UltraCap and Y-doped nanocrystalline HA was 1.62 and 1.60, respectively. Y mol % in Ca was 0.7 mol% and 0.4 mol% for UltraCap and nanocrystalline HA powders, respectively.

Mean particle sizes for UltraCap and nanocrystalline HA are listed in Table I. Equivalent spherical crystallite diameters calculated from the particle surface area was in the range of 6 to 8 μm for UltraCap powders, while that of the nanocrystalline HA powders was in the range of 31 to 32 nm. HA particles are presumed to increase crystallite size and particle size during sintering. On the contrary, HA treated hydrothermally inhibited grain growth and maintained a nanometer size crystal. Both powders were agglomerated since there was a large discrepancy between the mean particle size and crystallite size.

The UltraCap and nanocrystalline HA powders were agglomerated and had a broad size distribution (Fig. 2), which is consistent with the results obtained by the laser particle size analyzer. It was observed that UltraCap powders were more dense due to sintering and the particle shape was irregular. Fig. 2 (B) shows that small particles (about 1 μm) adhered to larger particles.

Table I: Characteristics of the synthesized HA powders

	UltraCap	Nanocrystalline	Y-doped UltraCap	Y-doped nanocrystalline
BET surface area (m²/g)	0.26	60.8	0.29	62.2
Equivalent spherical particle diameter (nm)	7400	31	6600	31
Mean particle size(μm) (Median (μm))	120 (169)	6.0 (4.8)	110 (140)	3.3 (4.4)

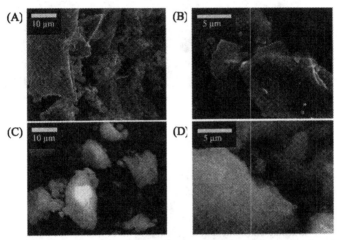

Fig. 2: SEM micrographs of (A) and (B) UltraCap HA powders as well as of (C) and (D) nanocrystalline HA powders.

Ion Tite™ coatings

Chemical phase was maintained after depositing the original UltraCap and nanocrystalline HA powders on titanium according to XRD (data not shown). Surface morphology of UltraCap and nanocrystalline HA coatings appeared to be similar to the original particle morphology (Fig. 3). AFM images also revealed that the crystallite size of nanocrystalline HA powders was maintained even after depositing on titanium substrates (Fig. 4). Surface area of nanocrystalline HA coatings was the highest, while RMS values were not significantly different between coatings (Table II). These results attribute to the fact that agglomeration size of the original powders for both UltraCap and nanocrystalline were in the micron scale and the surface area of the original nanocrystalline HA powders were significantly higher than that for the UltraCap powders.

67

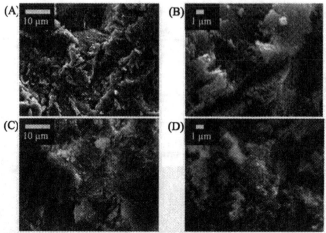

Fig. 3: SEM micrographs of IonTite™ titanium coatings of: (A) and (B) UltraCap HA as well as of (C) and (D) nanocrystalline HA.

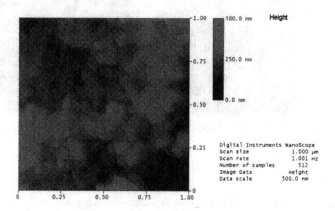

Fig. 4: AFM image of IonTite™ nanocrystalline HA coating.

Table II: Surface area of IonTite ™ titanium coatings

	Plasma-sprayed	UltraCap	Nanocrystalline
Surface area (μm^2)	1.68 ± 0.34	1.60 ± 0.35	2.79 ± 0.39
RMS (nm)	132.79 ± 59.44	85.26 ± 42.82	129.23 ± 33.64

* Projected area was 1 μm^2

Osteoblast functions – adhesion and calcium deposition

Osteoblast adhesion improved the most on undoped IonTite™ nanocrystalline compared to plasma-sprayed HA coatings (p<0.1) (Fig. 5). All HA coatings except Y-doped UltraCap statistically increased osteoblast adhesion more than uncoated titanium (p<0.05). Moreover, calcium deposition was the highest on Y-doped nanocrystalline HA coatings of all the samples tested; specifically, it was higher than plasma-sprayed HA coatings, undoped and Y-doped UltraCap HA coatings (p<0.05) (Fig. 6). Mean calcium concentration deposited by osteoblasts was greater on Y-doped nanocrystalline HA coatings than on undoped ones by more than two times, although statistical difference was not confirmed.

CONCLUSIONS

Undoped and Y-doped HA nanocrystalline powders were synthesized through a hydrothermal treatment process. Nanophase HA powders were deposited on titanium utilizing IonTite™ processing which did not change the chemistry and crystallite size of the starting HA powders. Osteoblast adhesion improved on nanocrystalline IonTite™ coatings compared to traditionally-used plasma-sprayed HA coatings. IonTite™ Y-doped nanocrystalline HA coatings stimulated osteoblasts to deposit more calcium compared to plasma-sprayed and UltraCap HA coatings. These results encourage further studies on nanophase IonTite™ HA coatings for orthopedic applications.

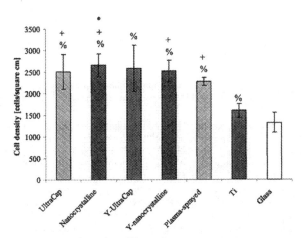

Fig. 5: Osteoblast adhesion on IonTite™ coatings of UltraCap HA, Y-doped UltraCap HA, nanocrystalline HA, Y-doped nanocrystalline HA, plasma-sprayed HA, uncoated titanium (Ti), and glass. Values are mean ± SEM; n=3; * p<0.1 (compared to plasma-sprayed coatings), + p<0.05 (compared to uncoated titanium), and % p<0.05 (compared to glass).

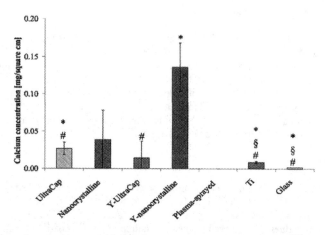

Fig. 6: Calcium deposited by osteoblasts after 21 days of culture on IonTite™ coatings of UltraCap HA, Y-doped UltraCap HA, nanocrystalline HA, Y-doped nanocrystalline HA, plasma-sprayed HA, uncoated Ti, and glass. Values are mean ± SEM; n=1 (in triplicate); * p<0.05 (compared to plasma-sprayed HA coatings), § p<0.05 (compared to UltraCap HA), and # p<0.05 (compared to IonTite™ coatings of Y-doped nanocrystalline HA).

ACKNOWLEDGEMENTS

This work was supported in part by NIH Phase SBIR Grant 1 R43 AR049657-01A.

REFERENCES

[1] C. Chang, J. Wu, D. Mao, C. Ding, "Mechanical and histological evaluations of hydroxyapatite-coated and noncoated Ti6Al4V implants in tibia bone," *J. Biomed. Mater. Res.*, 56, 17-23 (2001).

[2] L. Sun, C. Berndt, K. Gross, A. Kucuk, "Material fundamentals and clinical performance of plasma-sprayed hydroxyapatite coatings: A review," *J. Biomed. Mater. Res.*, 58, 570-592 (2001).

[3] A. Porter, L. Hobbs, V. Rosen, M. Spector, "The ultrastructure of the plasma-sprayed hydroxyapatite-bone interface predisposing," *Biomaterials*, 23, 725-733 (2002).

[4] L. Cleries, J. Fernandez-Pradas, G. Sardin, J. Morenza, "Dissolution behavior of calcium phosphate coatings," *Biomaterials*, 19, 1483-1487 (1998).

[5] T. Webster, C. Ergun, R. Doremus, R. Siegel, and R. Bizios, "Enhanced functions of osteoblasts on nanophase ceramics," *Biomaterials*, 21, 1803-1810 (2000).

[6] T. Webster, C. Ergun, R. Doremus, R. Bizios, "Hydroxyapatite with substituted magnesium, zinc, cadmium, and yttrium. II. Mechanisms of osteoblast adhesion," *J. Biomed. Mater. Res.*, 59, 312-317 (2002).

[7] J. Bronzino, "Biomedical Engineering Handbook," *CRC Press*, pp274-706 (1995)

[8] K. Ioku, M. Yoshimura, "Stochiometric apatite fine single crystals by hydrothermal synthesis," *Phosphorus Research Bulletin*, 1, 15-20 (1991).

A COMPARATIVE EVALUATION OF ORTHOPAEDIC CEMENTS IN HUMAN WHOLE BLOOD

N. Axén, N.-O. Ahnfelt, T. Persson, L. Hermansson,
Doxa AB
Axel Johanssons gata 4-6
SE-751 26 Uppsala
Sweden

J. Sanchez, R. Larsson
Uppsala University
The Rudbeck Laboratory
SE-751 85 Uppsala
Sweden

ABSTRACT

The clotting behaviour of some orthopaedic cements has been evaluated in a model system with human whole blood. The interest in the hemocompatibility of bone cements has increased because of claimed risks for adverse coagulation caused by material entering the blood system as the materials are injected into orthopaedic cavities.

This work investigates four orthopaedic cements: A polymethylmetacrylate (traditional bone cement) and three calcium-based ceramic cements, using a close circuit Chandler loop model with the inner surfaces of the PVC tubing coated by heparin. The model exposes the test materials to fresh human whole blood. A special procedure was developed to evaluate solidifying pastes in the Chandler loop model. This procedure covers a section of the inner wall of the tubing with a thin layer of non-cured cement paste. Thereafter the tubing is filled with fresh whole blood and the loop is closed. The loops are rotated at 32 rpm in a 37°C water bath for 60 minutes. The cements are curing in contact with the flowing blood.

After the incubation, the blood and the materials surfaces were investigated with special attention to clotting reactions. Blood samples were collected and supplemented with EDTA for cell count analysis. Blood from the loops was centrifuged to generate plasma for analysis of TAT, C3a and TCC complement marker. It is concluded that the clotting behaviour of the PMMA and the Ca-aluminate materials is considerably lower than that of the calcium phosphate and sulphate materials in these tests.

INTRODUCTION

The increasing incidence of osteoporosis worldwide drives a strong development of new bone substitute materials and methods for minimally invasive fracture treatment. This trend is illustrated in the spreading clinical acceptance of the vertebroplasty and kyphoplasty spinal treatments, involving stabilisation of collapsed vertebrae by percutaneous transpedicular cement injection into the damaged spinal bodies [1].

Traditional acrylic bone cement (PMMA) is well established and used both for implant fixation, fracture treatment (including vertebrae compression fractures) as well as for cranio-maxillo-facial applications [1, 2]. Also several new types of orthopaedic cements or bone void fillers used as pastes and which cure *in vivo* are commercially available. Most of these products are ceramic and based on various calcium salts such as calcium phosphates or calcium sulphates, but also new ceramic cements are entering the market, such as the calcium aluminate material evaluated in this work [3-6].

Calcium salt based cements have many advantages to the polymer materials; the absence of organic monomers, lower exothermal curing reactions, a higher degree of bioactivity and for some of these materials a bio-resorption over time are favourable factors.

The interest in the hemocompatibility of bone cement and the injectable ceramic pastes has increased because of claimed risks for adverse coagulation as material may enter the blood system unexpectedly during injection into vertebral bodies, e.g. during vertebroplasty or kyphoplasty. This work investigates the clotting behaviour of some common bone substitute materials, which are used for or are candidates for vertebro- and kyphoplasty.

The process that leads to thrombosis formation as blood contacts an artificial surface depends on a range of factors coupled to the material and its surface characteristics, the rheology and the biological aspects commencing with the initial protein adsorption [7-8].

MATERIALS AND METHODS

Materials

Four commercial or experimental versions of orthopaedic cements - *in vivo* curing bone graft pastes - have been evaluated in this work, one polymer and three ceramic materials:

A commercially available polymethylmethacrylate (PMMA) bone cement. The cement is free from antibiotics and radio-opacity additives. The material was prepared according to the instructions.

A commercially available calcium phosphate based cement or bone void filler for fracture treatments. The material was prepared by mixing powder and liquid in the recommended proportions by hand in a bowl.

An experimental calcium sulphate paste. The material is based on pure calcium sulphate hemi-hydrate powder (Aldrich) and de-ionised water in the powder to liquid ratio 2:1 by weight.

A calcium aluminate based cement (Doxa Biocement). This material is intended for vertebrae compression fractures and currently in clinical trials. The product is described in detail in [5].

Experimental procedure

The materials were studied in contact with fresh whole blood collected from healthy donors without addition of anticoagulant. In total, three different blood donors were used. The cements were mixed just before positioned in the test equipment and cured in blood contact.

The materials were tested with a closed circuit Chandler loop model [7]. In this test, the blood circulation is simulated with a rotating loop of PVC tubing with a total length 500 mm and an inner diameter of 4 mm, making a total inner volume of 6.3 ml. The inner surface of the PVC tubing is coated with an immobilized functional heparin. After application of the test materials and 4.5 ml (leaving a small air pocket facilitating the flow) of fresh blood, the closed tubing loops are rotated for one hour at 32 rpm in a 37°C water bath. Thereafter the blood and test materials are collected and investigated.

The cements/pastes were applied to the inner tubing wall by an especially developed procedure, aiming at covering a short section of the tubing wall with a thin layer of paste. The procedure aims at spreading the cement cylindrically towards the tubing wall to get sufficient lumen, and causing minimal rheological disturbance during the incubation.

First, the tubing is cut in two pieces, one short (ca. 80 mm), in which the test material is applied, and one longer filled with the donor blood. The shorter section is filled from one end with 0.02-0.04 ml of paste using a syringe and a short 10 gauge needle. Thereafter the tubing is compressed with the fingers to smear the paste over the tubing inner wall. Hence a ca.15 mm long zone of the inner tubing wall is covered with a thin layer of non-cured paste. The piece of tubing with cement is then rotated at 5000 rpm to further slung the paste towards the tubing wall.

Each tubing loop was rotated for 60 minutes at 32 rpm in a 37°C water bath. After the incubation, the blood was extracted from the loop, supplemented with EDTA, and the materials surfaces were investigated with special attention to clotting reactions and fibrin formation.

Three test runs were performed with different blood donors. Each test run included two loops for each material and a control loop making a total of 9 loops per test run. For each loop, a blood sample was immediately analysed in a cell counter. A plasma sample was prepared by centrifugation and immediately frozen to −70°C. A range of tests were performed on the blood/plasma, see table 1. Also the donor blood was tested, without having passed the loop test (baseline).

Blood evaluation

The collected blood samples and the test materials surfaces were first evaluated with respect to clotting by the eye. The blood was positioned on a cloth to find possible clots or contaminations from the test materials. The general clotting reactions were photographed.

Blood samples were also collected for centrifugation to separate plasma from cells. The plasma was evaluated with respect to TAT (thrombin-antithrombin complex), C3a and TCC (terminal complex of complement) markers. The set of tests was selected with respect to the ISO 10993-4 standard for blood contacting devices.

As for the platelet count, TAT, C3a and TCC collected values are presented according to the principle: For each material and test type, the highest and the lowest values are removed, and the span width of the remaining values is provided.

Table 1. Blood evaluation tests.

Evaluation	Test type
Thrombosis	Ocular investigation, photography, thrombosis counts
Coagulation	Visible clotting
Platelets	Platelet counts (for thrombocyte activation)
Immunology	C3a, TCC for complement activation
Fibrinogen	Thrombin-antithrombin complex (TAT)

RESULTS

Materials surfaces

The materials behaved differently to the exposure to flowing blood during curing. The PMMA and the Doxa Biocement both seemed relatively unaffected by the blood test. Both materials remained clad to the tubing inner wall. The calcium phosphate and the calcium sulphate materials, however, had interacted strongly with the blood and formed blood-ceramic material mixtures that largely filled the lumen of the tubing. The remaining material was found collected in large lumps of thrombosis, which were found loose in the tubing, see Fig. 1.

Filtration of the blood revealed some residues from the Doxa Biocement and PMMA materials. Blood having been exposed to the Doxa Biocement contained small amounts of small grains or crystals. The PMMA left small amounts of fragments as strings of polymer in the blood.

In summary the PMMA and the Ca-aluminate based Doxa Biocement caused only a mild clotting behaviour, whereas the calcium sulphate and phosphate materials caused a very strong clotting reaction in these tests.

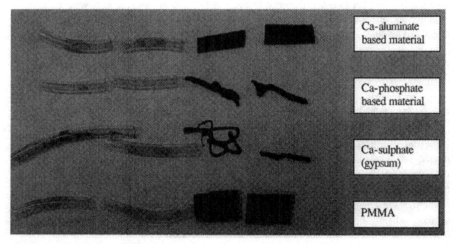

Figure 1. Macroscopic outcome of the blood exposure to the test materials.

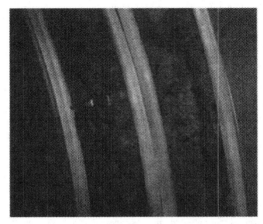

Figure 2. The surface of the Doxa Biocement after incubation with blood.

Blood response

The clotting behaviour and platelet count of the Doxa Biocement and the PMMA were comparable, both showing very low clotting and maintaining high platelet count numbers, around 200 for both materials. The calcium phosphate and sulphate materials were strikingly different but both showed very strong clotting as well as low platelet counts, around 10, see Table 2.

After centrifugation, no or low signs of haemolysis were found for the Doxa Biocement and the PMMA, whereas the calcium phosphate and calcium sulphate materials produced some haemolysis, as revealed by a reddish colour of the plasma.

Hence both the visual indications and the platelet count result show that calcium phosphate and calcium sulphate materials produced strong clotting in these tests. The Doxa Biocement and PMMA both produce low clotting and the platelet count values are comparable to the control.

Table 2. Blood analysis after termination of the Chandler loop test.

Material	Platelet count	C3a (µg/ml)	TCC (AU/ml)	TAT (pmol/ml)
Doxa Biocement (Ca-aluminate based)	198-271	628-751	137-481	7.4-29.0
Ca-phosphate based	8-37	-	-	-
Calcium sulphate	4-14	-	-	-
PMMA	217-308	303-444	109-126	3.4-11.6
Control	144-304	152-317	30-39	2.6-40.9
Baseline	229-335	34-417	8.5-12	2.1-3.6

Also, the control loops did not induce any complement activation as reflected by formation of C3a. Both PMMA and Doxa Biocement induced moderate activation of the C3a component of the complement. The Doxa Biocement displayed somewhat higher tendency in this respect. Also the control had a slightly increased TCC value compared to the baseline.

The formation of the soluble terminal complex of complement, TCC was slightly elevated in the control loops. This has previously been shown to be linked to the presence of an air interface. The TCC data obtained from the loops with PMMA and Doxa Biocement were consistent with the data obtained on C3a.

DISCUSSION

Both PMMA and Doxa might induce some complement activation, following undesirable flow into the blood circulation, but these cements should not be expected to cause thrombotic complications in view of their very low tendency to induce activation of the coagulation system. In view of the overall inflammatory response associated with the orthopaedic procedures the tendency to activate complement factors is not considered to be harmful at a considerable level but should stimulate research aiming at formulation of new cements with less potential to activate complement factors. It should be emphasised that complement may not only be activated in blood but also in tissue.

The concentration of thrombin-antithrombin complex (TAT) was largely unaffected by the test procedure for the PMMA and Doxa Biocement materials. For one blood donor, the concentration of thrombin-antithrombin complex (TAT) rose by a factor of ten in the control loops, i.e. the loops that had no deposition of cement, as well as for the loops with test material. This elevation reflects slight activation due to rotation of blood in the presence of an air interface. There was no further generation of TAT. The cements based on phosphate and sulphate formulations, however, induced clotting to such an extent that macroscopic clots were formed and all platelets were consumed. According to previous experience, such vigorous activation produces TAT levels well beyond 1000 pmol/ml, and therefore, these samples were not analysed.

It might be hypothesised that calcium phosphate and calcium sulphate have a higher solubility than calcium aluminate, thereby producing elevated calcium levels in the blood that could trigger the coagulation cascade. It can also be hypothesized that calcium aluminate is more reactive than the other calcium salts in producing its end product, calcium aluminate hydrate and hydroxylapatite at the surface. The consequence of this, in comparison to the other calcium salts, might be that calcium levels are not elevated when blood is exposed to calcium aluminate. A surface reaction with phosphate, calcium and calcium aluminate hydrate will lead to the formation of hydroxylapatite, and possibly to a reduction of calcium in the blood [9].

In order to prove this reaction scheme, it is necessary to follow the calcium levels as a function of time during the rapid curing of the calcium aluminate hydrate. An analysis of the dissolution and precipitation characteristics and reaction kinetics of the different cement systems, and the resulting effects on pH and calcium levels is believed to be central in the understanding of the different behaviours on blood clotting.

CONCLUSIONS

In conclusion, the clotting behaviour of the Ca-aluminate based Doxa Biocement and the PMMA was much lower than for the calcium phosphate and calcium sulphate materials. Both materials also maintained high platelet count numbers, whereas the calcium phosphate and sulphate materials showed very low platelet counts.

The PMMA and Doxa Biocement materials should not be expected to cause thrombotic complications in view of their very low tendency to induce activation of the coagulation system.

REFERENCES

[1]S. M. Kenny, M. Buggy, Bone cements and fillers: A review, Journal of Materials Science: Materials in Medicine 14 (2003) 923-938.

[2]L. E. Jasper, H. Deramond, J. M. Mathis, S. M. Belkoff, Materials properties of various cements for use with vertebroplasty, Journal of Materials Science: Materials in Medicine 13 (2002) 1-5.

[3]G. Baroud, M. Bohner, P. Heini and T. Steffen, Injection biomechanics of bone cements used in vertebroplasty, Bio-Medical Materials and Engineering 00 (2004) 1-18.

[4]M. L. Roemhildt, T. D. Mcgee, S. D. Wagner, Novel calcium phosphate composite bone cement: strength and bonding properties, Journal of Materials Science: Materials in Medicine 14 (2003) 137-141.

[5]N. Axén, T. Persson, K. Björklund, H. Engqvist and L. Hermansson, An injectable bone void filler cement based on Ca-aluminate, Key Engineering Materials Vols. 254-256 (2004) 265-268.

[6]J. Husband, C. Cassidy, C. Leinberry, M. S. Cohen, J. Jupiter, Multicenter clinical trial of Norian SRS versus conventional therapy in treatment of distal radius fractures. Trans. Amer. Acad. Orthop. Surg. (Abst): 403, 1997.

[7]J. Andersson, J. Sanchez, K. Nilsson Ekdahl, G. Elgue, B. Nilsson, R. Larsson, Optimal heparin surface concentration and antithrombin binding capacity as evaluated with human non-anticoagulated blood in vitro, J. Biomed. Mater. Res. A. Vol 67 (2) 458-66 (2003)

[8]J. Andersson, K. Nilsson Ekdahl, R. Larsson, U. R. Nilsson, B. Nilsson, C3 adsorbed to a polymer surface can form an initiating alternative pathway convertase, I. Immunol. Vol 168 (11) 5786-91 (2002)

[9]H. Engqvist, J-E. Schultz-Walz, J. Loof, G. A. Botton, D. Maye , M. W. Pfaneuf, N-O.Ahnfelt , L. Hermansson "Chemical and Biological Integration of a Mouldable Bioactive Ceramic Material capable of forming Apatite in vivo in Teeth". Biomaterials vol 25 (2004) pp. 2781-2787

SELF-SETTING ORTHOPEDIC CEMENT COMPOSITIONS BASED ON CaHPO₄ ADDITIONS TO CALCIUM SULPHATE

SELF-SETTING ORTHOPEDIC CEMENT COMPOSITIONS BASED ON $CaHPO_4$ ADDITIONS TO CALCIUM SULPHATE

J. N. Swaintek[1], C. J. Han[1], A. C. Tas[2], and S. B. Bhaduri[2]

[1] S. C. Governor's School for Science and Mathematics, Hartsville, SC 29550
[2] School of Materials Science and Engineering, Clemson University, Clemson, SC 29634

ABSTRACT

Calcium sulphate-based cements depend on the setting reaction between water and calcium sulphate hemihydrate (CSH, $CaSO_4 \cdot 1/2H_2O$) to form calcium sulphate dihydrate (CSD, Gypsum, $CaSO_4 \cdot 2H_2O$) as the reaction product. Rapid formation of gypsum needles provides the resultant material its initial, dry cohesive strength. Such weak cements, in the form of granules, pellets or cement pastes, are commercially available as bone defect fillers in clinical orthopedic applications. However, pure calcium sulphate cements rapidly deteriorate in aqueous solutions and crumble into a powder. Calcium sulphate cements are also not able to maintain their dry strength when soaked in human blood plasma or synthetic body fluids at 37°C. Additions of calcium phosphate powders (5 to 33 wt%), such as $CaHPO_4$ (DCPA, dicalcium phosphate anhydrous, monetite) to CSH were found to significantly increase the wet mechanical integrity of these new cements. *In vitro* apatite-inducing ability of pure gypsum cements and the DCPA-doped calcium sulphate cements were compared by soaking those in a tris-buffered, 27 mM HCO_3^- containing synthetic body fluid (SBF) solution for 1 week. While the gypsum cement samples were not able to form any carbonated apatitic calcium phosphates on their surfaces, DCPA-doped cement samples were covered with a thick layer of carbonated, apatitic calcium phosphate. Moreover, the DCPA-doped gypsum cements kept their initial mechanical strength after 1 week of soaking in the SBF solution. Samples of pure gypsum cements, on the other hand, simply disintegrated into loose powders during the same SBF soaking.

INTRODUCTION

Since the 19th century, calcium sulphates have been in use as non-load-bearing bone grafts or dental implants, and Dreessmann [1] even noted that some of the ancient Egyptian mummies were found to have dental fillings made out of calcium sulphate. Calcium sulphate has shown excellent biocompatibility [2] and in 1980 Coetzee [3] concluded that it should be regarded as a good bone graft substitute, comparable to autograft, in defects in the skull and facial bone [4]. With the addition of water, calcium sulphate hemihydrate (CSH, $CaSO_4 \cdot 1/2\ H_2O$) converts into calcium sulphate dihydrate (CSD, Gypsum, $CaSO_4 \cdot 2H_2O$) as the end product. This material, after its FDA approval, is widely used as a bone defect-filling material [5-8].

However, calcium sulphate cements have few points of concerns: (*i*) its rapid passive dissolution and *in vivo* resorption even before the host bone has had the time to grow into the defect area [4], (*ii*) its known cytotoxic effect: its dissolution leads to acidic microenvironment responsible for local inflammatory processes at the site of implantation in human bone [3] (Inflammatory tissue was found to disappear after 60 days in bone, but to remain in soft tissue implantation sites of white New Zealand rabbits [9].), (*iii*) pure calcium sulphate is not able to maintain its initial dry strength when soaked in water or synthetic body fluids; it disintegrates, and (iv) pure calcium sulphate does not have the property of osteoconductivity. To address these concerns and to improve the properties of calcium sulphate cements, calcium phosphate additions (β-TCP, α-TCP, HA and $Ca(H_2PO_4)_2 \cdot H_2O$) have been investigated [4, 10-12].

Compressive strength of pure calcium sulphate dihydrate cements shows a significant variation over the range of 1 to 15 MPa depending on the liquid-to-powder (L/P) ratio employed in preparing the pastes [13, 14]. With an increase in the L/P ratio, strength rapidly deteriorates due to the creation of remnant porosity in the final cement bodies. Fernandez *et al.* [15] reported that α-TCP (which is by itself a self-setting cement powder) additions to CSD can be used to increase the compressive strength values to above 20 MPa. Sato *et al.* [16] showed that osteoconductivity in calcium sulphate cements and new bone formation could be improved when at least 50% of calcium sulphate was substituted by HA powders.

To the best of our knowledge, addition of $CaHPO_4$ to CSH powders has not been studied before. The present paper reports the mixing of $CaHPO_4$ (from 5 to 33 wt%) with CSH powders. The resultant cements were biphasic mixtures of CSD and $CaHPO_4$, with a significant carbonated, biomimetic apatite-inducing ability upon soaking in SBF solutions [17] for 1 week at 37°C.

EXPERIMENTAL PROCEDURE

Pure calcium sulphate pellets were prepared as follows: 12.0 g of CSH powder *(99.8%, Aldrich, Milwaukee, WI)* was kneaded, by using an agate pestle, with 4.8 mL of deionized water (i.e., L/P=0.4) in an agate mortar for 3 minutes. The formed paste was then transferred into a stainless steel die of a diameter of 2.54 cm. The paste in the die was then uniaxially pressed at a pressure of 1.78 kg/mm² for 8 minutes. The recovered pellets were dried overnight at 37°C in air.

$CaHPO_4$ *(>99%, J. T. Baker, Phillipsburg, NJ)*-containing calcium sulphate cement pellets were prepared and studied at three compositions: 5, 10 and 33 wt% $CaHPO_4$. For the preparation of, for instance, 33 wt% $CaHPO_4$ samples, 4.0 g of $CaHPO_4$ and 8.0 g of CSH powders were first mixed, under 6 mL of high purity ethyl alcohol, in an agate mortar. These ethanol-mixed powders were then dried in a glass Petri-dish at 37°C, overnight. Dried powders were then mixed with water in an agate mortar at the L/P ratio of 0.4. The remainder of the processing and pellet formation was exactly the same as with those described above for pure calcium sulphate. The obtained pellets were of 2.54 cm diameter and about 4 mm height.

Samples were characterized by using an X-ray diffractometer, XRD, (XDS 2000, Scintag, Sunnyvale, CA) operated at 40 kV and 30 mA with mono-chromated Cu K_α radiation, by using FTIR, Fourier Transformed Infrared Spectroscopy, (Nicolet 550, Thermo-Nicolet, Woburn, MA), and by FESEM, field-emission scanning electron microscopy (S-4700, Hitachi, Tokyo, Japan). Mechanical testing (i.e., the load versus displacement curves) was performed using an Universal Testing Machine (Instron, Phoenix 20K, MTI. Roswell, GA). The pellets were placed between self-levelling plates and compressed at 1 mm/min.

Pellets of pure CSD and CaHPO$_4$-doped CSD cements were both soaked in 100 mL of synthetic body fluid (SBF) solutions for 1 week at 37°C to examine and compare their apatite-inducing ability. SBF solutions were prepared as described in Table I below [17]. SBF solutions were replenished with fresh solutions at every 48 hours during 1 week of soaking.

Table I.   Preparation of SBF solution (1 L)

Order	Reagent	Weight (g)	Ion	Human Plasma(mM)	SBF (mM)
1	NaCl	6.547	Na$^+$	142	142
2	NaHCO$_3$	2.268	Cl$^-$	103	125
3	KCl	0.373	HCO$_3^-$	27	27
4	Na$_2$HPO$_4$·2H$_2$O	0.178	K$^+$	5	5
5	MgCl$_2$·6H$_2$O	0.305	Mg^{2+}	1.5	1.5
6	CaCl$_2$·2H$_2$O	0.368	Ca^{2+}	2.5	2.5
7	Na$_2$SO$_4$	0.071	HPO$_4^{2-}$	1	1
8	(CH$_2$OH)$_3$CNH$_2$	6.057	SO$_4^{2-}$	0.5	0.5

RESULTS AND DISCUSSION

Figure 1 depicts the XRD data of the starting powders of CaSO$_4$·1/2H$_2$O (CSH). The XRD data for the CaSO$_4$·2H$_2$O (CSD) powders formed after mixing CSH powders with water is also given in Figure 1 (top trace). CSH reacts with water according to the following reaction

$$CaSO_4 \cdot 1/2 \ H_2O + 3/2 \ H_2O \rightarrow CaSO_4 \cdot 2 \ H_2O \qquad (1).$$

The XRD data reported in the top trace of Fig. 1 was gathered from the ground powders of formed CSD pellets described above. In a similar fashion, Figure 2 shows the FTIR traces of pure CSH and pure CSD powders. These XRD and FTIR served as controls prior to the start of CaHPO$_4$-doping in CSH powders.

The XRD data for 5, 10, and 33% CaHPO$_4$-doped calcium sulphate samples (Figure 3) showed that all three samples readily allowed the conversion of CSH into CSD. In other words, even the increasing presence of monetite did not inhibit the CSH to CSD transformation. Figure 4 depicted the FTIR traces of 5, 10, and 33% CaHPO$_4$-doped calcium sulphate samples. The FTIR traces were all quite similar to that of CSD. The shoulder-like bands over the range of 1400 to 1300 cm^{-1}, appeared in the top FTIR trace (i.e., 33wt% CaHPO$_4$) of Fig. 4, was due to the HPO$_4^{2-}$ groups. Characteristic IR bands of pure CaHPO$_4$ were observed at 2803, 2326, 2110, 1630, 1399, 1345, 1124, 1060, 986, 884, 555, 535 cm^{-1} (data not shown here).

SEM morphology of pure CSH powders was shown in Figure 5(a). CSH powders consisted of particles in the range of 5 to 40 μm. On the other hand, the CSH to CSD transformation totally consumed these large particles, and resulted in the formation of CSD whiskers, as shown in Figure 5(b). These interlocking, intermingling whiskers or needles provide the dry strength to the CSD samples.

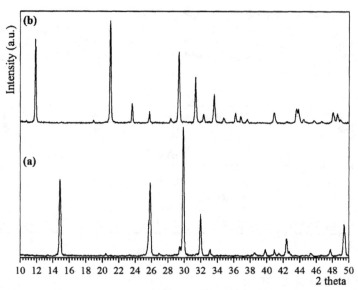

Fig. 1   XRD traces of (*a*) pure CSH and (*b*) pure CSD

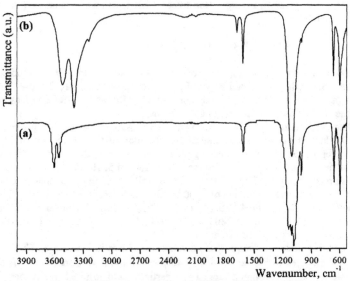

Fig. 2   FTIR traces of (*a*) pure CSH and (*b*) pure CSD

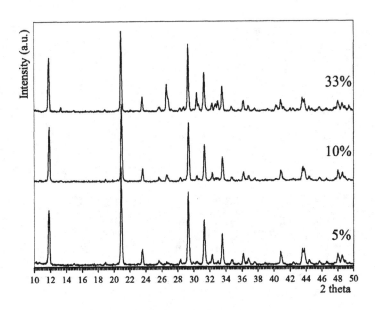

Fig. 3 XRD traces of 5, 10, and 33% CaHPO₄ in CSD cement

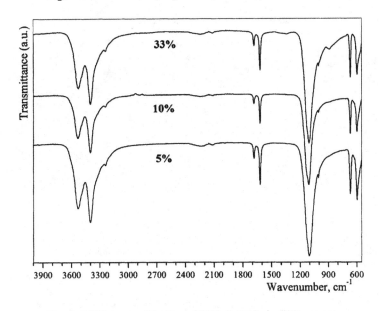

Fig. 4 FTIR traces of 5, 10, and 33% CaHPO₄ in CSD cement

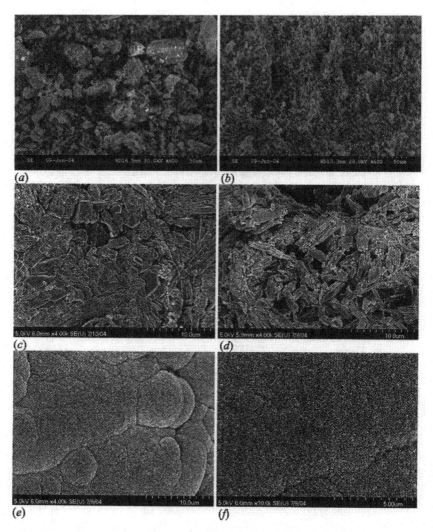

Fig. 5 FESEM micrographs of (a) initial CSH powders, (b) pure CSD cement, (c) 10% CaHPO₄ in CSD cement, (d) 33% CaHPO₄ in CSD cement, (e) 33% CaHPO₄ in CSD sample after SBF-soaking (*low mag*), (f) close up view of the sample in (e).

The FESEM micrograph of Figure 5 (c) showed the surface of 10% CaHPO₄-containing CSD sample, which was soaked in SBF for 1 week. It was apparent that these samples (like the pure CSD and 5% CaHPO₄-containing samples) did not have any apatite-inducing ability. This was the also same morphology of the 10% CaHPO₄-

containing samples prior to the SBF soaking. Figure 5(d) depicted the FESEM morphology of a 33% CaHPO$_4$-containing sample prior to the SBF soaking. Figures 5(e) and 5(f) showed the 33% CaHPO$_4$ sample after 1 week of SBF soaking at low and high magnifications. This was the typical morphology of biomimetically (i.e., in SBF, at 37°C and pH 7.4) coated, carbonated, nanoapatites. When one went to higher magnifications, the nanotexture of this material became visible, which comprised interlocking nanocrystals of carbonated, calcium- and OH ion-deficient, apatitic calcium phosphate [18]. Therefore, such 33% CaHPO$_4$-doped gypsum cements revealed, for the first time, a significant apatite-inducing ability, which was not present in pure gypsum.

Moreover, with the addition of CaHPO$_4$ into calcium sulphate, the load-displacement curves (under compression, on cylindrical samples having a surface area of 507 mm^2) improved considerably with respect to those of pure calcium sulphate cements, as shown in Figure 6. 33% CaHPO$_4$-containing samples had a higher load-bearing ability in comparison to pure CSD.

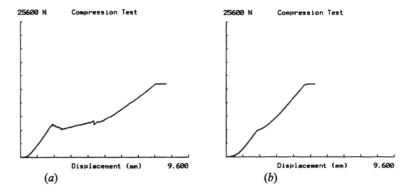

Fig. 6. Load-displacement curves of (a) pure CSD, and (b) 33% CaHPO$_4$-CSD sample

CONCLUSIONS

CaHPO$_4$ powder additions (5, 10, and 33wt%) were performed in initially CSH powder matrices to test (1) self-setting cement behavior, (2) apatite-inducing ability in SBF solutions, and (3) mechanical integrity of the resultant bodies. While 5 and 10wt% doped CSD bodies did not show any apatite-inducing ability, 33wt% CaHPO$_4$-containing samples displayed a remarkable degree of apatite formation on the sample surfaces after 1 week of SBF soaking at 37°C. 33% CaHPO$_4$ doped samples (prior to SBF soaking) also showed an improved load-bearing ability in comparison to pure CSD cements. These CaHPO$_4$ doped calcium sulphate cements are, thus, expected to show *in vivo* osteoconductivity.

ACKNOWLEDGMENTS

Dr. S. B. Bhaduri and Dr. A. C. Tas are grateful to the South Carolina Governor's School for Science and Mathematics in making possible for the two bright high school students (J. N. Swaintek and C. J. Han) to spend 4 weeks of their summer (June 2004) as research interns at Clemson University.

REFERENCES

[1] H. Dreessmann, "Ueber Knochenplombierung bei Hohlenformigen Defekten des Knochens," *Beitr. Klin. Chir.*, 9, 804-810 (1892).

[2] L. F. Peltier, "The Use of Plaster of Paris to Fill Defects in Bone," *Clin. Orthop.*, 21, 1-31 (1961).

[3] A. S. Coetzee, "Regeneration of Bone in the Presence of Calcium Sulfate," *Arch. Otolaryngo.*, 106, 405-409 (1980).

[4] M. Nilsson, L. Wielanek, J. S. Wang, K. E. Tanner, and L. Lidgren, "Factors Influencing the Compressive Strength of an Injectable Calcium Sulfate-Hydroxyapatite Cement," *J. Mater. Sci. Mater. M.*, 14, 399-404 (2003).

[5] Smith & Nephew; *http://ortho.smithnephew.com/us/Standard.asp?NodeId=3287*

[6] Wright Medical Technology; *http://www.wmt.com/Physicians/Products/Biologics/OSTEOSETBoneGraftSubstitute.asp*

[7] Orthogen Corp.; *http://www.orthogencorp.com/pages/886726/index.htm*

[8] Lifecore Biomedical; *http://www.lifecore.com/products/capset.asp*

[9] A. Strauss, "Lokaler Antibiotikumtraeger aus Kalziumsulfat: Vertraeglichkeit im Gewebe und Pharmakokinetik der angewendeten Antibiotika nach Implantation in Kaninchen." In: *Biomaterialien in der Medizin*. Giessen, Germany: Koehler; 1999. pp. 105–108.

[10] C. Liang, Z. Li, D. Yang, Y. Li, Z. Yang, W. W. Lu, "Synthesis of Calcium Phosphate/Calcium Sulphate Powder," Mater. Chem. Phys., 88, 285–289 (2004).

[11] M. Bohner, "New Hydraulic Cements Based on Alpha-Tricalcium Phosphate-Calcium Sulfate Dihydrate Mixtures," *Biomaterials*, 25, 741-749 (2004).

[12] M. Nilsson, E. Fernandez, S. Sarda, L. Lidgren, and J. A. Planell, "Characterization of a Novel Calcium Phosphate/Sulphate Bone Cement," *J. Biomed. Mater. Res.*, 61, 600-607 (2002).

[13] A. Gisep, S. Kugler, D. Wahl, and B. Rahn, "Mechanical Characterization of a Bone Defect Model Filled with Ceramic Cements," *J. Mater. Sci. Mater. M.*, 15, 1065-1071 (2004).

[14] N. B. Singh, "The Activation Effect of K_2SO_4 on the Hydration of Gypsum Anhydrite, $CaSO_4$, *J. Am. Ceram. Soc.*, 88, 196–201 (2005).

[15] E. Fernandez, M. D. Vlada, M. M. Gela, J. Lopez, R. Torres, J. V. Cauich, and M. Bohner, "Modulation of Porosity in Apatitic cements by the use of α-Tricalcium Phosphate-Calcium Sulphate Dihydrate Mixtures," *Biomaterials*, 26, 3395-3404 (2005).

[16] S. Sato, T. Koshino, and T. Saito, "Osteogenic Response of Rabbit Tibia to Hydroxyapatite Particle-Plaster of Paris Mixture," *Biomaterials*, 19, 1895-1900 (1998).

[17] D. Bayraktar and A. C. Tas, "Chemical Preparation of Carbonated Calcium Hydroxyapatite Powders at 37°C in Urea-containing Synthetic Body Fluids," *J. Eur. Ceram. Soc.*, 19, 2573-2579 (1999).

[18] A. C. Tas and S. B. Bhaduri, "Rapid Coating of Ti6Al4V at Room Temperature with a Calcium Phosphate Solution Similar to 10× Simulated Body Fluid," *J. Mater. Res.*, 19, 2742-2749 (2004).

ADHESIVE STRENGTH OF THE APATITE LAYER FORMED ON TiO₂ NANOPARTICLES/HIGH DENSITY POLYETHYLENE COMPOSITES

Masami Hashimoto, Hiroaki Takadama and Mineo Mizuno
Japan Fine Ceramics Center
2-4-1 Mutsuno, Atsuta-ku, Nagoya, 456-8587, Japan

Tadashi Kokubo
Research Institute for Science and Technology, Chubu University
1200 Matsumoto-cho, Kasugai, 487-8501, Japan

ABSTRACT

The adhesive strength was studied between bone-like apatite layer and composites consisted of high density polyethylene (HDPE) and TiO_2 (hereafter TiO_2/HDPE). Apatite formed on the TiO_2/HDPE composite after 3 days soaking in a simulated body fluid (SBF), when the composite was first subjected to polishing with 10 % HNO_3 or HCl, and then to soaking in the SBF. The adhesive strength was measured using a modified ASTM C-333 standard, in which a tensile stress was applied to the substrate. Measured strength of the apatite layer was compared with HAPEX® which is commercially available hydroxyapatite-polyethylene composite. The TiO_2/HDPE composites demonstrated to have much greater as three times adhesive strength (maximum 4.5±0.5 MPa) as compared with HAPEX® (1.4±0.5 MPa), when filler contents of TiO_2/HDPE and HAPEX® were 50 and 40 vol%, respectively. Because the matrix material in both TiO_2/HDPE and HAPEX® is the same, the adhesive strength is attributed to the increase in the number of the apatite nuclei on the TiO_2 nanoparticles. The higher adhesive strength of the apatite layer formed on the TiO_2/HDPE composite suggests that this composite might also be utilized in applications involving higher levels of load bearing such as vertebra.

INTRODUCTION

Since the discovery of Bioglass® in the early 1970s [1], various kinds of bioactive ceramics including sintered hydroxyapatite [2] and glass-ceramic A-W containing crystalline apatite and wollastonite ($CaO \cdot SiO_2$) [3] were developed and clinically used as artificial middle-ear bones, maxillofacial implants, bone fillers, iliac crests, vertebrae, intervertebral disc, etc. Among them, glass-ceramic A-W shows both a high mechanical strength of 200 MPa [4, 5] and high bioactivity [6]. The material can be used such as vertebrae and intervertebral disc which bears under rather loaded sites [7]. Highly loaded bones such as tibial and femoral bones, however, can not be replaced even with this glass-ceramic, since its fracture toughness is lower and its elastic modulus is higher than those of the natural bone. For these purposes, metallic implants such as titanium metal and its alloy previously subjected to alkali and heat treatment [8] or alkali, water and heat treatment [9] have been developed. They have, however, much higher elastic moduli than that of the natural bone. This is a critical problem, since high elastic modulus of the materials may induce resorption of the surrounding bone because of their stress shielding. Therefore, highly bioactive materials with mechanical properties analogous to those of the natural bone are desirable for development.

Hydroxyapatite particle reinforced high-density polyethylene (HDPE) composite (HAPEX®) has been developed since the early 1980s as an analogue for bone [10]. It is already clinically used as artificial middle-ear bone. Some of the mechanical properties of the HAPEX®, such as the tensile strength, have already have been found to be desirable for its use in the body [11-13]. But fracture toughness and elastic modulus of HAPEX® are lower than those of living bone. And also, glass-ceramic A-W reinforced HDPE has been developed since 1998 [14, 15]. The bioactivity of this composite is higher than that of HAPEX®; however, the mechanical strengths of this composite are lower than those of HAPEX®.

TiO_2 nanoparticles reinforced HDPE composite (hereafter TiO_2/HDPE) was previously shown to exhibit bending strength and young's modulus analogous to those of natural bone as well as bioactivity [16, 17]. The bending strength and young's modulus were previously shown to vary from almost 28 to 54 MPa and 1.4 to 7.6 GPa, respectively, depending on the TiO_2 content. These composites form a bone-like apatite layer on their surface in a simulated body fluid (SBF: K^+ 5.0, Na^+ 142, Mg^{2+} 1.5, Ca^{2+} 2.5, Cl^- 147.8, HCO_3^- 4.2, HPO_4^{2-} 1.0, SO_4^{2-} 0.5mM), which is comparable to that formed in vivo when adhesive to living bone. The apatite was formed on the TiO_2/HDPE composite after 7 days soaking in the SBF.

In this study, TiO_2/HDPE composite with the filler content of 50 vol% was first subjected to polishing with various kinds of solution such as H_2O, HCl or HNO_3, and then to soaking in SBF. Effect of the kind of solution in polishing on the apatite forming ability was studied. And then, to develop an appreciation of the adhesive strength of the composites with living bone, the tensile strength of the bond between the apatite layer formed in the SBF and the surfaces of the TiO_2/HDPE composite was examined as an analog of the in vivo system. The results for the TiO$_2$/HDPE were compared with that of HAPEX®, which had a filler content of 40 vol%.

MATERIALS AND METHODS

Materials; Solvents and reagents, all of special reagent grade, were used without further purification. An anatase-type TiO$_2$ nanopowder was manufactured by Ishihara Sangyo Kaisha, Ltd., Mie, Japan. The median particle size of TiO$_2$ powders was 535 nm.

HDPE (Japan Polyolefins Co., Ltd., Tokyo, Japan) has following number-average molecular weight; Mn in 1.21x10^4, weight-average molecular weight; Mw in 7.67x10^4 and z-average molecular weight; Mz in 47.6x10^4, Mw/Mn in 6.35 and Mz/Mw in 6.20. Melt flowing rate (MFR) of this polyethylene is 8.

Preparation of TiO$_2$/HDPE Composites; The manufacturing process of the TiO$_2$/HDPE involved kneading and compression moulding. The filler content was set at 50 vol% because this composite shows both bioactivity and mechanical properties analogous to those of human cortical bone [17]. This composite was denoted as TiO$_2$/HDPE–50. HDPE was dried at 80 °C for 8hrs and then kneaded at 210 °C in a batch kneader PBV 0.3 (Irie shokai, Ltd., Tokyo, Japan). TiO$_2$ particles were added slowly into the melted HDPE with kneading at 210 °C in air. After adding TiO$_2$, TiO$_2$/HDPE compound was kneaded with 25 rpm rotation speed for 30 min. The obtained compounds were molded at 230 °C for 1 hour and then hot-pressed in air at 2.5 MPa.

Characterization of TiO$_2$/HDPE composites

Fourier transform infrared spectroscopy; The spectra of the TiO$_2$/HDPE-50 composite after polishing using SiC paper under pouring H$_2$O, 10 %HCl or HNO$_3$ were analyzed with a FT-IR spectrophotometer FT/IR-550 (JASCO Co., Osaka, Japan). The polished composite surface, 1 mg finely divided, were ground and dispersed in a matrix of KBr (150 mg), followed by compression to consolidate the formation of the pellet. FT-IR spectra were obtained by the KBr pellet method in the wavenumber range of 400 to 4000 cm^{-1} with a resolution of 4 cm^{-1}.

Apatite formation; It has been revealed that materials, which form a bone-like apatite on its surface in SBF, form the apatite even in the living body and bond to living bone through the apatite layer [18]. For TiO$_2$/HDPE composite, specimens of 10 mm x 10 mm x 4 mm in size were cut, polished with a 220 grit silicon carbide paper for 5 min under pouring H$_2$O, 10 %HCl or HNO$_3$, washed with distilled water and dried at room temperature. SBF with ion concentrations nearly equal to those of human blood plasma was prepared by dissolving reagents NaCl, NaHCO$_3$, KCl, K$_2$HPO$_4$¥3H$_2$O, MgCl$_2$¥6H$_2$O, CaCl$_2$ and Na$_2$SO$_4$ (Nacalai tesque, Inc. Kyoto, Japan) in distilled water and buffered at pH7.4 at 36.5 °C with (CH$_2$OH)$_3$CNH$_2$ and 1M HCl (Nacalai tesque, Inc. Kyoto, Japan). The specimens were soaked in 30ml of SBF at 36.5 °C. After various periods, the specimens were removed from the fluids, washed moderately with ion-exchanged distilled water, and dried at room temperature for 1 day. Their surfaces were analyzed by thin-film X-ray diffraction (TF-XRD) with RINT Model 2000 (Rigaku Denki Co. Ltd., Tokyo, Japan). The morphology of the surface layer of the composites was observed by FE-SEM with an S-800 Model (Hitachi Ltd., Tokyo, Japan) after coating a thin Au film.

Adhesive strength measurement; The adhesive strength was measured using a method used by Juhasz et. al, which is a modified ASTM C-633 method. Ten samples of each composite type were attached one at a time to 10 mm x 10 mm x 15 mm stainless steel jigs using Rapid-type Araldite glue (Nichiban Co. Ltd., Kobe, Japan). After the glue had dried fully, the samples were tensile tested using a tensile testing machine, SERVOPULSER (Shimadzu Co. Ltd., Kyoto, Japan) operating at a cross-head speed of 1 mm/min. The load was applied normal to the sample test surfaces (10 mm x 10 mm). The average adhesive strength was calculated for each composite type using the load at fracture and the sample surface area.

RESULTS AND DISCUSSIONS

Figure 1 show the characteristic spectra of TiO_2/HDPE-50 polished with H_2O, 10% HCl or HNO_3 solution. Comparing the composites spectra, differences in the shape and position of the bands can be observed. The most significant bands of the TiO_2/HDPE-50 polished with 10% HCl or HNO_3 solution studied and the assignment for each one is the broad adsorption band between 550-653 and 436-495 cm^{-1} assigned to Ti-O and Ti-O-Ti linkages in TiO_2 nanoparticles [19-20]. On the other hand, in the TiO_2/HDPE-50 polished with H_2O, it is necessary to highlight the no characteristic band assigned to Ti-O-Ti. This result indicates that the HDPE covered on the TiO_2 nanoparticles was selectively removed by polishing with only 10% HCl or HNO_3 solution.

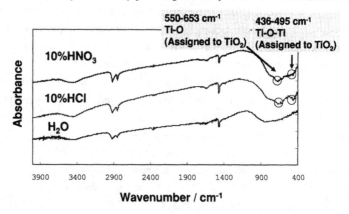

Figure 1   FT-IR spectra of TiO_2/HDPE-50 polished with H_2O, 10 % HCl or HNO_3.

90

Figures 2 and 3 show SEM photographs and thin-film X-ray diffraction patterns of the surfaces of TiO_2/HDPE-50 polished with H_2O, 10% HCl or HNO_3 solution, then subjected to soaking in the SBF for various periods, respectively. Figure 2 shows that there formed a little product on the TiO_2/HDPE-50 polished with H_2O, even after the soaking in the SBF for 7 days. This product was assigned to the apatite by thin-film X-ray diffraction pattern (Figure 3). For polishing with 10% HCl or HNO_3 solution, the apatite was formed after 3 days soaking in the SBF. Especially, a dense and uniform apatite layer was formed on the HNO_3-polished TiO_2/HDPE-50. It has been found that Ti-OH group on the exposed TiO_2 nanoparticles provides preferred sites for apatite nucleation.

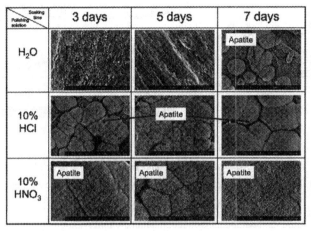

Figure 2   FE-SEM images of the surfaces of TiO_2/HDPE-50 polished with H_2O, 10% HCl or HNO_3 solution, then subjected to soaking in the SBF for various periods.

The nominal adhesive strengths between the apatite layers and the composites are shown in Figure 4. The adhesive strength of TiO_2/HDPE-50 polished with H_2O and the HAPEX® were similar, apart from the TiO_2/HDPE-50 polished with 10% HCl solution, which had a tensile strength more than double that of HAPEX®. Comparison of the two TiO_2/HDPE-50 composites polished with 10% HCl or HNO_3 solution showed that the apatite layers on TiO_2/HDPE-50 polished with HNO_3 possessed slightly greater adhesive strength. This increase is well correlated with the induction period reduction. Therefore, it is attributed to the increase in the number of apatite nuclei as well as to the increase in their adhesive strength to the composite.

Analysis of the SEM trace of the TiO_2/HDPE-50 (Figure 5), after tensile testing, showed the apatite layer to be completely removed, but at the same time, TiO_2 particles were not removed from the HDPE matrix.

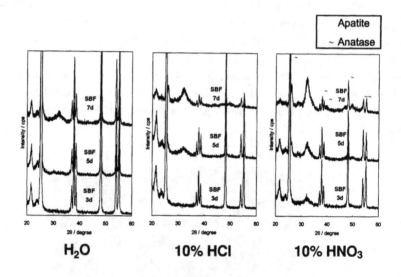

Figure 3   Thin-film X-ray diffraction patterns of the surfaces of TiO$_2$/HDPE-50 polished with H$_2$O, 10% HCl or HNO$_3$ solution, then subjected to soaking in the SBF for various periods

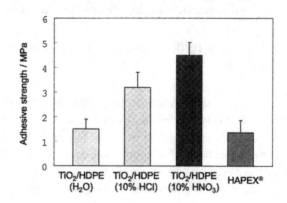

Figure 4   Adhesive strength of apatite layer formed on TiO$_2$/HDPE and HAPEX® for comparison.

92

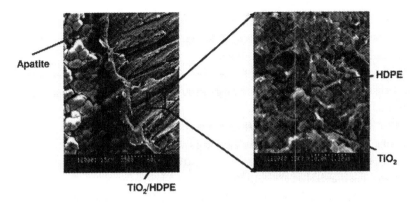

Figure 5 FE-SEM images of TiO₂/HDPE-50 (polished with HNO₃) surfaces after adhesive strength testing

CONCLUSIONS

HCl or HNO_3 polishing selectively removed HDPE with which TiO_2 nanoparticles were covered. TiO_2 exposure on TiO_2/HDPE-50 significantly enhanced apatite nucleation on TiO_2/HDPE-50 and increased adhesive strength of the apatite layer to the composite. The present method is found to give highly bioactive materials, which can be applied to bone-repairing materials.

ACKNOLEDGEMENT

This work is in part supported by the National Research & Development Programs for Medical and Welfare apparatus entrusted from the New Energy and Industrial Technology Development Organization (NEDO) to the Japan Fine Ceramics Center.

REFERENCES

[1] L. L. Hench, RJ. Splinter, WC. Allen and T. K Jr. Greenlee, J. Biomed. Mater. Res. 2 117-41 (1971)

[2] M. Jarcho, JL. Kay, RH. Gumaer and HP. Drobeck, J. Bioeng. 1 79-92 (1977)

[3] T. Kokubo, M. Shigematsu, Y. Nagashima, M. Tashiro, T. Nakamura, T. Yamamuro and S. Higashi, Bull. Inst. Chem. Res. Kyoto Univ. 60 260-8 (1982)

[4] T. Kokubo, S. Ito, M. Shigematsu, S. Sakka and T. Yamamuro, J. Mat. Sci. 20 2001-4 (1985)

[5] T. Kokubo, S. Ito, M. Shigematsu, S. Sakka and T. Yamamuro, J. Mat. Sci. 22 4067-70 (1987)

6   K. Ono, T. Yamamuro, T. Nakamura and T. Kokubo, Biomaterials 11 265-271 (1990)

7   T. Yamamuro, J. Shikata, H. Okumura, T. Kitsugi, Y. Kakutani, T. Matsui and T. Kokubo, J. Bone and Joint Surg., 72-B 889-893 (1990)

8   HM. Kim, F. Miyaji, T. Kokubo and T. Nakamura, J. Biomed. Mater. Res. 32 409-17 (1996).

9   M. Uchida, HM. Kim, T. Kokubo, S. Fujibayashi and T. Nakamura, J. Biomed. Mater. Res., 63(5) 522-30 (2002)

10 W. Bonfield, MD. Grynpas, AE. Tully, J. Bowman and J. Abram, Biomaterials 2 185-6 (1981)

11 M. Wang, R. Joseph and W. Bonfield, Biomaterials 19 2357-66 (1998)

12 M Wang and W. Bonfield, Biomaterials 22 1311-20 (2001)

13 M. Wang, Biomaterials 24 2133-51 (2003)

14 JA. Juhasz, SM. Best, W. Bonfield, M. Kawashita, N. Miyata, T. Kokubo and T. Nakamura, J. Mater. Sci: Mater. In Med. 14 489-95 (2003)

15 JA. Juhasz, SM. Best, R. Brooks. M. Kawashita, N. Miyata, T. Kokubo, T. Nakamura and W. Bonfield, Biomaterials 25 949-955 (2004)

16 H. Takadama, M. Hashimoto, Y. Takigawa, M. Mizuno, Y. Yasutomi and T. Kokubo, Key Engineering Materials 254-256 569-572 (2004)

17 M. Hashimoto, H. Takadama, M. Mizuno and T. Kokubo, J. Mater. Sci: Mater. in Med. submitted.

18 T. Nakamura, M. Neo and T. Kokubo, in "Mineralization in Natural and Synthetic Biomaterials" ed. by P. Li, P. Calvert, T. Kokubo, R. Levy and C. Sheid (Materials Research Society, Warrendale, PA, 2000) p.15-25.

19 C. Xie, Z. Xu, Q. Yang, B. Xue, Y. Du and J. Zhang, Materials Science & Engineering B 112 34-41 (2004)

20 T. Bezrodna, G. Puchkovska, V. Shymanovska, J. Baran and H. Ratajczak, Journal of Molecular Structure, in press.

EFFECT OF REINFORCEMENTS ON PROPERTIES OF SELF-SETTING CALCIUM PHOSPHATE CEMENT

N. C. Bhorkar and W. M. Kriven
Department of Materials Science and Engineering
University of Illinois at Urbana-Champaign
Urbana, IL 61801, USA.

ABSTRACT
Calcium phosphate cement makes a suitable material to be used for craniofacial and orthopedic repairs because of its biocompatibility, bioresorbability and quick setting properties. These materials have an advantage over polymers and metals in that they are non-toxic and lightweight. However, the low strength of calcium phosphate prohibits its use in stress bearing bone joints. This study investigates the improvements in mechanical properties of calcium phosphate cement by reinforcements. Monocalcium phosphate monohydrate (Aldrich Chemicals, Milwaukee, WI) and β-tricalcium phosphate, synthesized by the organic steric entrapment method, were the primary components of the cement. Upon setting the cement transformed into brushite. β-TCP and cement after setting, were analyzed by X-ray analysis, EDS. Two different types of reinforcements were used, namely, (a) chopped hydroxyapatite (HA) fibers, and (b) borosilicate glass spheres. Rod shaped particles (less than 500 μm in length) were made by chopping the sintered HA fibers (about 115-120 μm in diameter). Borosilicate glass spheres, prepared by glass melting methods, had diameters in the range of 25 to 70 μm. Samples were prepared with different weight fractions of the reinforcements, and tested both in flexure and compression mode. Effects of size, shape, and weight fraction of the inclusions on the strength of the biocomposite are discussed.

INTRODUCTION

Calcium phosphate cements (CPC) form a smooth paste upon mixing with water and the cementitious reaction produces a hard mass.[1,2] β-tricalcium phosphate (β-TCP, β-$Ca_3(PO_4)_2$) and monocalcium phosphate monohydrate (MCPM, $Ca(H_2PO_4)_2 \cdot H_2O$) upon mixing with water show cement-like behavior. Dissolution of MCPM followed by the precipitation of small crystals of dicalcium phosphate dihydrate (DCPD) entangling the β-TCP particles helps to gain strength as the cement sets. However, β-TCP–MCPM cement set very rapidly (in less than 30 s) and their diametral strength is rather low (<1 MPa).[3,4] It has been shown that addition of calcium phosphate hemihydrate (CSH, $CaSO_4 \cdot \frac{1}{2}H_2O$) increases the setting time as well as the strength of the hardened cement.[3] Little investigation has been carried out on the reinforcement of CPCs.[3] In this study, we investigated the effect of reinforcements, in particular, chopped hydroxyapatite (HA) fibers and borosilicate glass spheres on the mechanical strength of CPC.

EXPERIMENTAL PROCEDURE

Synthesis of β-tricalcium phosphate

β-tricalcium phosphate powder was synthesized by the organic steric entrapment method.[5,6] $Ca(NO_3)_2 \cdot 4H_2O$ (99 %, Sigma-Aldrich, Milwaukee, WI) and H_3PO_3 (99 %, Sigma-Aldrich, Milwaukee, WI), used as calcium and phosphorus source respectively, were dissolved in D.I. water. After complete dissolution, 5 wt % aqueous solution of polyvinyl alcohol was added.

The mixture was stirred and heated until the water was completely evaporated. The dried mass was then crushed in a porcelain mortar using a pestle. The powder was then calcined and crystallized at 850°C/1h and 1075°C/4h respectively followed by grinding in an alumina mortar.

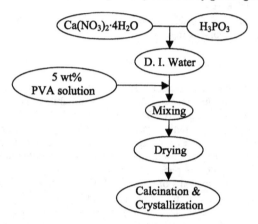

Fig. 1. Schematic of organic steric entrapment method used for synthesis of β-TCP powder

Hydroxyapatite (HA) fibers

Hydroxyapatite (HA) powder (extra pure, Sigma-Aldrich, Milwaukee, WI) was attrition milled using Szegvari Attritor System (Union Process, Akron, OH) for 1h at 50 % power and sieved to obtain particle size less than 32 μm. The powder was then mixed with cellulose (5 wt %) and the paste was extruded using a fiber extruder (Marksman V, Joyce Corp., Dayton, OH) through a die of diameter 150 μm. The fibers so obtained were sintered at 950°C for 2 h and then chopped manually using blade followed by sieving through 500 μm size sieve.

Cement formulation

β-TCP, MCPM (Sigma-Aldrich, St. Louis, MO) and CSH (Sigma-Aldrich, St. Louis, MO) were dry mixed manually with a spatula for about 15 s and then using an alumina mortar and pestle for about 30 s. The mixture was then mixed with cold D.I. water for about 30 s using a spatula. It has been shown that β-TCP-MCPM cement shows maximum strength at about 80-20 wt % mixture.[7] The various cement formulations made are shown in Table I. The cement was reinforced with chopped hydroxyapatite (HA) fibers and borosilicate glass spheres (prepared by glass melting method, Mo-Sci Specialty Products Corp., Rolla, MO). The mass content of reinforcements was varied from 5 to 30 wt %.

96

Table I. Cement powder formulation

Components	Pure Cement wt %	Compositions w/ reinforcements (wt %)							
		#1	#2	#3	#4	#5	#6	#7	#8
MCPM	16	16	16	16	16	16	16	16	16
β-TCP	64	64	64	64	64	64	64	64	64
CSH	20	20	20	20	20	20	20	20	20
Chopped HA fibers*	-	5	10	15	30	-	-	-	-
Borosilicate glass spheres*	-	-	-	-	-	5	10	15	30

Mechanical testing
The cement paste was filled in a cylindrical stainless steel mold of diameter 6 mm and height of 12 mm for compression testing. The dimensions of the mold for flexure testing were 2.3 mm × 2.3 mm × 25 mm. The samples were allowed to set for an hour and then removed from the mold to dry for 24 h ± 2 h at an estimated 50 % relative humidity. before testing. The load cell used in both tests was 1 kN and the loading rate was 20 mm/min and 0.5 mm/min for compression and flexure test respectively.

Characterization
The shape and size of the HA fiber and borosilicate glass spheres were characterized using a scanning electron microscope (SEM, Model S-4700, Hitachi, Osaka, Japan). The energy dispersive spectroscopy (EDS) was used to determine the Ca/P ratio in β-tricalcium phosphate. The phases of synthesized β-TCP and the cement after setting were verified using powder X-ray diffraction (Rigaku D-Max, Tokyo, Japan). Figure 2 shows the powder X-ray diffraction pattern for β-TCP powder synthesized by the organic steric entrapment method. SEM was also used to qualitatively determine any microporosity in the set cement.

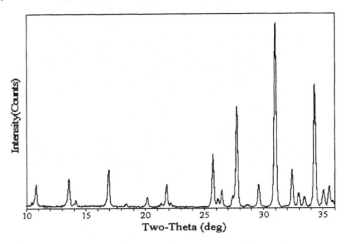

Fig. 2. Powder X-ray diffraction pattern of β-TCP powder

RESULTS AND DISCUSSION

Reinforcements

a) HA fibers: HA fibers extruded through a die of diameter 150 μm, showed shrinkage after sintering at 950°C/2h down to 115-120 μm. The fibers still showed some microporosity after sintering (Fig. 3). The HA grains were found to be of 2-3 μm in size as shown in Fig. 4.

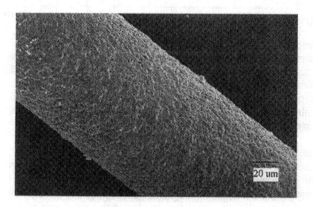

Fig. 3. Scanning electron micrograph of HA fiber (diameter ~ 115-120 μm)

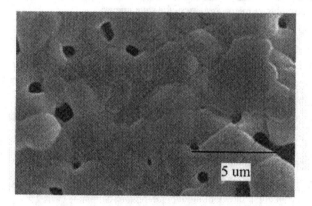

Fig. 4. Scanning electron micrograph showing grains inside HA fiber.

b) Borosilicate glass spheres: Commercial grade borosilicate glass spheres were also used as reinforcement. The manufacturer specified particle size was ≤ 74 μm and it was confirmed qualitatively using SEM (about 25-70 μm) as shown in Fig. 4.

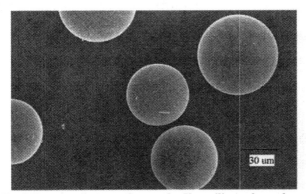

Fig. 4. Scanning electron micrograph of borosilicate glass spheres.

Cement

Synthesis of β-TCP using the organic steric entrapment method yielded very fine powder with high surface area. The SEM micrograph shows platelet-shaped β-TCP particles (Fig. 5). Smaller particle size with higher surface energy resulted in very reactive powder. It has been shown that addition of CSH helps to slow down the setting reaction through formation of needle-like crystals entangled within larger β-TCP particles.[1,3] The setting mechanism consists of the dissolution of MCPM followed by the gradual crystallization of DCPD throughout the paste. The hardening occurs as a result of precipitation of DCPD crystals, binding together remaining β-TCP particles.

Fig. 5. Scanning electron micrograph of β-TCP powder after calcination at 850°C/1h and crystallization at 1075°C/4h.

The transformation can be explained by the following reaction[1,7]:

$$Ca(H_2PO_4)_2 \cdot H_2O \ (s) + Ca_3(PO_4)_2 + 7H_2O \ \rightarrow \ 4CaHPO_4 \cdot 2H_2O \ (s)$$

X-ray diffraction pattern of the set cement (after 24 h ± 2 h), crushed into fine powder using an alumina mortar and pestle, indicated peaks of dicalcium phosphate dihydrate ($CaHPO_4 \cdot 2H_2O$) also known as brushite, along with peaks of remaining β-TCP (Fig. 6).

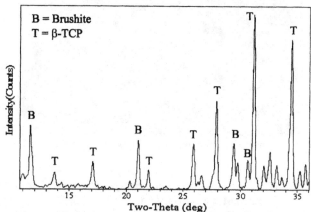

Fig. 6. Powder X-ray diffraction pattern of the cement (24 h after mixing) showing brushite peaks along with unreacted β-TCP.

The effect of chopped HA fiber and borosilicate glass sphere reinforcements on compressive strength of the set cement is shown in Fig. 7. Additions of chopped HA fibers up to 15 wt % improved the compressive strength of the cement by 28 %. The strength went down upon further addition of chopped HA fibers. On the other hand, addition of borosilicate glass showed almost no improvement in compressive strength. It seemed to have detrimental effect instead. Similar observations were made in case of flexure testing of the samples with varying content of reinforcements (Fig. 8). The chopped HA fibers increased the flexure strength by a maximum of 45 % upon addition of 15 % by weight. The strength declined on further addition of HA fibers. The borosilicate glass addition did not improve the flexure strength, but rather decreased it beyond 10 wt %.

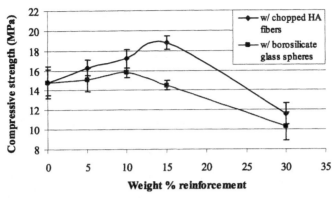

Fig. 7. Effect of reinforcements on compressive strength. Mean values plotted with error bars showing standard deviation (SD); n = 5.

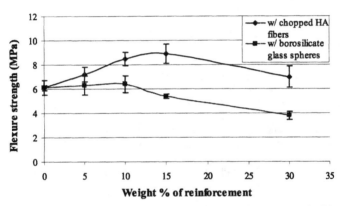

Fig. 8. Effect of reinforcements on flexure strength. Mean values plotted with error bars showing standard deviation (SD); n = 5.

The improvement in strength using chopped HA fibers could be due to fiber pullout effects and crack deflection. It is possible that spherically shaped borosilicate glass did not form a bond with the surrounding cement and thus failed to improve the strength. Further studies are needed to examine the fracture interface for debonding and crack deflection effects of reinforcements.

CONCLUSIONS

High yield β-TCP powder was successfully synthesized by the organic steric entrapment method at relatively low temperature. β-TCP-MCPM-CSH-water cement system formed DCPD

101

phase upon setting. Addition of rod-shaped HA particles by up to 15 wt % improved the compressive and flexure strength, but both declined upon further addition. A slight increase in strength was observed for the borosilicate glass reinforcement and maximum strength was seen at a 10 wt % addition. Further addition of glass spheres decreased the strength to lower than the strength of the pure cement without any reinforcements. The high aspect ratio HA rods had a positive effect on strength as compared to spherical-shaped, borosilicate glass.

AKNOWLEDGEMENTS

The authors thank Dr. P. Sarin and other colleagues in the laboratory for their help at various stages of this work.

FOOTNOTES
* Weight % of pure cement weight without any reinforcements.

REFERENCES

[1]K. Ohura, M. Bohner, P. Hardouin, J. Lemaitre, G. Pasquier, and B. Flautre, "Resorption of, and Bone Formation from, New β-Tricalcium Phosphate-Monocalcium Phosphate Cements: An *In Vivo* Study," *J. Biomed. Mat. Res.*, **30**, 193-200 (1996).

[2]M. Kamath and H. Varma, "Development of a Fully Injectable Phosphate Cement for Orthopedic and Dental Applications," *Bull. Mater. Sci.*, **26**, 415-422 (2003).

[3]A. Mirtchi, J. Lemaitre, and E. Munting, "Calcium Phosphate Cements: Action of Setting Regulators on the Properties of the β-Tricalcium Phosphate-Monocalcium Phosphate Cements," *Biomaterials*, **10**, 634-638 (1989).

[4]Y. Fukase, E. Eanes, S. Takagi, L. Chow and W. Brown, "Setting Reactions and Compressive Strength of Calcium Phosphate Cement," *J. Dent. Res.*, **69**, 1852-56 (1990).

[5]M. Gülgün, W. Kriven and M. Nguyen, "Processes for Preparing Mixed Metal Oxide Powders", *U.S. Patent* number 6,482,387, (2002).

[6]W. Kriven, M. Gülgün, M. Nguyen and D. Kim, "Synthesis of Oxide Powders via Polymeric Steric Entrapment," *Ceramic Transactions*, **108**, 99-110, (2000)

[7]A. Mirtchi, J. Lemaitre, and N. Terao, "Calcium Phosphate Cements: Study of the β-Tricalcium Phosphate-Monocalcium Phosphate System," *Biomaterials*, **10**, 475-480 (1989).

THE BIOACTIVITY OF PDMS-CaO-SiO$_2$ BASED HYBRID MATERIALS PREPARED BY THE ADDITION OF TRANSITION METAL ALKOXIDES

Manabu FUKUSHIMA[1†], Eiichi YASUDA[1], Hideki KITA[2], Masao SHIMIZU[1], Yasuto HOSHIKAWA[1] and Yasuhiro TANABE[1]

1)Materials & Structures Laboratory, Tokyo Institute of Technology, 4259, Nagatsuta, Midori-ku, Yokohama, Japan, 226-8503, JAPAN

2)National Institute of Advanced Industrial Science and Technology, 2266-98 Shimo-Shidami, Moriyama-ku, Nagoya, Aichi 463-8560, Japan

†Now with AIST> National Institute of Advanced Industrial Science and Technology, 2266-98 Shimo-Shidami, Moriyama-ku, Nagoya, Aichi 463-8560, Japan

ABSTRACT
The bioactivity of organic inorganic hybrid materials derived from polydimethylsiloxane, tetraethoxysilane, calcium nitrate tetrahydrate and a different amount of pentaethoxy-niobium/tantalum was examined. We investigated the relationship between the hydroxyapatite depositions on the hybrid's surface in a simulated body fluid (KOKUBO solution) and the molecular structure oh hybrids. The in vitro formation of hydroxyapatite was characterized by using a simulated body fluid and the molecular structure of the hybrid was characterized by fourier transformed infrared and silicon 29 solid-state magic angle spinning nuclear magnetic resonance. The addition of a different amount of transition metal alkoxides lead to a different amount of silanol in the obtained hybrid materials. Additionally, the releasing ability of calcium from hybrid into simulated body fluid was different among the obtained hybrid materials, corresponding to the amount of transition metal. The hybrid material with a ratio of 0/1 and 1/0=Nb/Ta, polydimethylsiloxane/tetraethoxysilane=0.1/1 and 0.1=(Ta+Nb)/tetraethoxysilane showed the higher apatite formation ability.

1. INTRODUCTION

Recently, bioactive organic inorganic hybrid materials have gathered the great interest because of lower elastic modulus and larger elongation than bioactive metals, alloy like titanium and alkali treated titanium, and ceramics such as bioglass [1-8]. These hybrid materials showed the similar elastic modulus with cancellous bone and the high bioactivity [4-8]. Generally, polydimethylsiloxane (PDMS) based organic inorganic hybrid materials have received a considerable attention, where, for their preparation, PDMS has been incorporated into CaO-SiO$_2$ and CaO-SiO$_2$-TiO$_2$, according to the following equation.

$$\text{TEOS} \xrightarrow[2hr]{\text{H}_2\text{O/HCl}} \text{PDMS} \xrightarrow[15hr]{} \begin{array}{c} \text{Ca(NO}_3)_2/\text{H}_2\text{O} \\ \text{or} \\ \text{Ca(NO}_3)_2/\text{H}_2\text{O/Ti(o-i-Pr)}_4 \end{array}$$

$$\xrightarrow[\text{1-2weeks/40°C}]{\text{Casting}} \xrightarrow[\text{3days/60°C}]{} \xrightarrow[\text{1day/150°C}]{} \text{Hybrid disk}$$

The siloxane molecular network in hybrid material is formed by the silicon alkoxides; tetraethoxysilane (TEOS). Calcium nitrate tetrahydrate was used as calcium source. For the evaluation of bioactivity, Kokubo solution (simulated body fluid) with an inorganic ion composition similar to that of human blood plasma was usually used [9]. The soaking of specimens into SBF in vitro experiments represents the reactions that may take place in vivo [10]. Osaka reported that the bioactivity of PDMS based hybrid material is affected by the calcium releasing ability from hybrid into SBF and silanol (Si-OH) concentrations of the hybrid [1-3]. Kokubo reported PDMS-CaO-SiO_2-TiO_2 hybrid with high bioactive showed the mechanical properties similar to human cancellous bone [4-7].

SiO_2 and TiO_2 from alkoxide were introduced for the forming of glass network. However, the introducing of another component except titania has not been reported sufficiently. Tantalum/niobium itself and their oxide have a preferred biocompatibility. The addition of tantalum/niobium alkoxide into TEOS has the possibility with the control of siloxane network and the silanol amount, because electronegativity and partial charge of silicon is different from that of transition metal.

In this paper, the preparation of PDMS based organic inorganic hybrid materials with containing multi component except titanium was attempted, and the relationship between the molecular structure and their apatite forming ability in SBF was investigated.

2. EXPERIMENTAL

2-1 Sample preparation

TEOS was stirred under the mixed solvent of isopropyl alcohol (IPA) and tetrahydrofuran (THF) with 35wt% HCl for catalyst and the distilled water for hydrolysis. After recognizing transparent solution of above mixture in approximately 2hr, PDMS (1000 in molecular weight) as organic source was added and mixed for 18hr. Then, the modified pentaethoxy-niobium/tantalum with acetylacetone and the ethanol-water solution of calcium nitrate tetrahydrate were added and stirred for 2hr, respectively. These procedures mentioned above were followed by Kokubo and his coworker's report [4-8]. The molar ratio of TEOS/IPA/THF/HCl/PDMS/Ca(NO_3)$_2$ was kept at 1/0.5/0.5/0.05/0.1/0.1, and (Ta+Nb)/TEOS was (0.1+0)/1, (0.05+0.05)/1 and (0+0.1)/1. The solutions were aged for 2days in Teflon shale at room temperature. They were dried at 40°C for 5days, 60°C for 2days and 150°C for 1day. The sample notations were represented by the ratio of added transition metal alkoxides (in mol%) as (A) Ta0Nb10, (B) Ta5Nb5 and (C) Ta10Nb0.

2-2 Characterizations

The structure of the obtained hybrid products was investigated by fourier transformed infrared (FT-IR) and silicon 29 solid-state magic angle spinning nuclear magnetic resonance (^{29}Si-MAS NMR) spectra. In FT-IR measurement, the powdered samples were mixed with KBr. The measurements were carried out by using SHIMADZU FTIR-8600PC. ^{29}Si-MAS NMR spectra were obtained using JEOL JNM-EXCALIBUR-270 MHz. Polydimethylsilane was used as an external standard. The sample spinning speed was approximately 6 kHz. All the spectra were recorded with a repetition time of 60s. In average, approximately 700 scans were carried out for obtaining a good S/N ratio.

2-3 Evaluation of Bioactivity

Simulated body fluid (SBF) was used for the evaluation of bioactivity, which was pH=7.40 and had an inorganic ion composition similar to that of human blood plasma (Na^+142.0, K^+5.0,

$Mg^{2+}1.5$, $Ca^{2+}2.5$, $Cl^-148.8$, $HCO_3^-4.2$, $HPO_4^{2-}1.0$ and $SO_4^{2-}0.5mM$). The specimens were soaked in 40ml of SBF at 36.5°C for various periods, rinsed with distilled water and then dried at room temperature. The surface of the products before and after soaking into SBF was analyzed with an X-ray diffract meter (XRD; RINT2000) and was observed with an environmental scanning electron microscope (E-SEM; Nikon E-SEM 2700). The changing of calcium concentration for calcium in SBF before and after soaking was investigated with inductively coupled plasma (ICP). The average value of two measurements was used.

3. RESULTS AND DISCUSSIONS
3-1 Structural characterization

Figure 1 shows the ^{29}Si-MAS NMR spectra of the obtained products. From spectra, two distinct peaks are due to D unit; $SiO_2(CH_3)_2$ and Q unit; $SiO_n(OH)_{4-n}$ detected at -15~21ppm and -100~-109ppm, respectively, where "n" means the number of siloxane bonding; n=0-3 [2, 11]. In this case, Q unit is divided into Q_3; $SiO_3(OH)$ and Q_4; SiO_4 structures, corresponding to the number of siloxane bonding. Silanol group (Si-OH) has an important factor for bioactivity. The amount of silanol increased in the order of Ta0Nb10≈Ta5Nb5<Ta10Nb0, calculated by the area of deconvoluted peak. In addition, the peak due to D unit in Ta5Nb5 product was detected at a slightly higher field than that of other hybrids. This means the copolymerization of PDMS itself, corresponding to chain structure and cyclic structure from PDMS [2, 11]. In fact, from the calculation of peak area of D unit, PDMS without cross linking to TEOS and pentaethoxy tantalum/niobium remained in the order of Ta5Nb5> Ta0Nb10>Ta10Nb0. This may increase the hydrophobic property.

FT-IR spectra of the products are illustrated in Fig.2. The absorption bands at 790 cm^{-1} due to νSi-CH_3 and at 1050 cm^{-1} due to $\nu(Si$-O-$Si)$ are detected [12], the band around 850 cm^{-1} is due to TEOS-PDMS copolymerization and the peak around 950 cm^{-1} is corresponding to Si-O-M (M=Ta and /or Nb, and H). In addition, the phenomenon that this band shifts to higher wave number was observed, with increasing tantalum content. This suggests the increasing amount of Si-OH, which is consistent with NMR result.

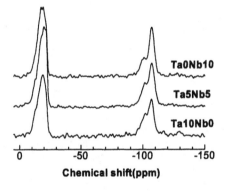

Figure 1 ^{29}Si-MAS NMR spectra of products

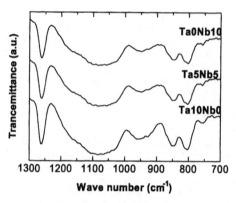

Figure 2 FT-IR spectra of products

3-2   The apatite formation in SBF
Table 1 illustrates the hydroxyapatite deposition period of the products before and after soaking into SBF for various periods up to 5 days, which were confirmed by XRD. The hydroxyapatite on the surface of all products after soaking into SBF was formed within 1day.

Table 1 The apatite deposition periods of products in SBF. These were confirmed by XRD.

	The apatite deposition period in SBF						
	0	6h	15h	1d	2d	3d	5d
Ta0Nb10	x	x	o	o	o	o	o
Ta5Nb5	x	x	x	o	o	o	o
Ta10Nb0	x	x	o	o	o	o	o

The SEM micrographs of the surfaces of the products before and after soaking into SBF for various periods are shown in Fig.3. These show the formation of hydroxyapatite on the surface hybrids. The hybrid's structural factors for the apatite deposition on the surface of hybrids in SBF are associated with the amount of hydroxyl group such as Si-OH, Nb-OH and Ta-OH and the releasing of calcium from the product into SBF [1-3]. Thus, the changing of calcium concentration of SBF was investigated, shown in Fig.4. The increasing of calcium after SBF soaking of 1 day was observed for all hybrids. This is due to the releasing from products into SBF. However, the releasing rate of Ta5Nb5 product is lower than other hybrids. From the result of Si-NMR spectra, it was found that PDMS leads to the more predominant copolymerization of PDMS itself than the cross linking with TEOS and pentaethoxy-niobium/tantalum, which suggests the hydrophobic property. The lower releasing rate of calcium from the molecular structure of Ta5Nb5 product into SBF may be connected with the hydrophobic property.

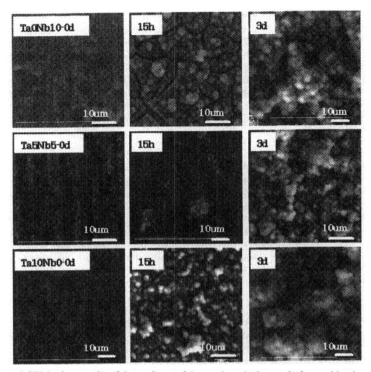

Figure 3 SEM micrographs of the surfaces of the products before and after soaking into SBF

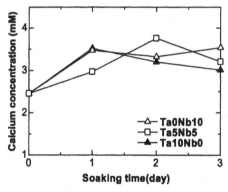

Fig.4 The changing of calcium concentration of SBF

In addition to the releasing rate of calcium, the silanol amount is important for the formation of apatite nucleation. As seen from Si-NMR results, the silanol amount between Ta5Nb5 and Ta0Nb10 products showed the similar value. The calcium releasing rate from Ta0Nb10 product into SBF is larger than that of Ta5Nb5 product. Therefore, it is thought that the apatite forming ability of Ta0Nb10 product in SBF was higher than that of Ta5Nb5 product.

In this paper, the fabrication of PDMS based organic inorganic hybrid materials with containing tantalum and niobium, and their apatite forming ability in SBF were investigated. The addition of transition metal alkoxide resulted in various silanol amounts and different calcium releasing rate into SBF, suggesting the controlled molecular structure to form the apatite on the surface of hybrid material in SBF.

4. CONCLUSION

The preparation of PDMS-CaO-SiO$_2$ based hybrids prepared by the different amount of pentaethoxy-niobium/tantalum and their apatite formation ability in SBF were investigated. The obtained hybrid materials showed various silanol amounts and a different calcium releasing rate into SBF. Although the hybrid's structure derived from 1/1=Nb/Ta, PDMS/TEOS=0.1/1 and 0.1=(Ta+Nb)/TEOS showed the similar silanol amount with that of the product from 1/0=Nb/Ta, PDMS/TEOS=0.1/1 and 0.1=(Ta+Nb)/TEOS, the apatite formation of former product in SBF was slower because of the lower calcium releasing rate into SBF. It is found that the addition of transition metal alkoxide into polymer and siloxane is usefully workable for the apatite forming ability in SBF, namely which can control silanol amount in the structure of hybrid and the calcium releasing ability.

REFERENCES
(1)   K.Tsuru, C.Ohtsuki, A.Osaka, T.Iwamoto and J.D.Mackenzie, "Bioactivity of sol–gel derived organically modified silicates Part I: In vitro examination", J.Mater.Sci.Mater.Med., 8, 3, 157-161 (1997).
(2)   T.Yabuta, E.O.Bescher, J.D.Mackenzie, K.Tsuru, S.Hayakawa, and A.Osaka, "Synthesis of PDMS-based porous materials for biomedical applications",J.Sol-Gel Sci.Tech., 26, 1219-1222 (2003).
(3)   K.Tsuru, Y.Aburatani, T.Yabuta, S.Hayakawa, C.Ohtsuki and A.Osaka, "Synthesis and in vitro behavior of organically modified silicate containing Ca ions", J.Sol-Gel Sci.Tech., 21, 89-96 (2001).
(4)   Q.Chen, N.Miyata, T.Kokubo and T.Nakamura, "Bioactivity and mechanical properties of PDMS-modified CaO–SiO$_2$–TiO$_2$ hybrids prepared by sol-gel process", J.Biomed.Mater.Res, 51, 605–611 (2000).
(5)   Q.Chen, M.Kamitakahara, N.Miyata, T.Kokubo and T.Nakamura, "Preparation of bioactive PDMS-modified CaO-SiO$_2$-TiO$_2$ hybrids by the sol-gel method", J.Sol-Gel Sci.Tech., 19, 1-3, 101-105 (2000).
(6)   Q.Chen, N.Miyata, T.Kokubo and T.Nakamura, "Effect of heat treatment on bioactivity and mechanical properties of PDMS-modified CaO-SiO$_2$-TiO$_2$ hybrids via sol-gel process", J.Mater.Sci.Mater.Med., 12, 6, 515-522 (2001).
(7)   Q.Chen, N.Miyata, T.Kokubo and T.Nakamura, "Bioactivity and mechanical properties of poly(dimethylsiloxane)-modified calcia-silica hybrids with added titania", J.Am.Ceram.Soc., 86, 5, 806-810 (2003).

(8)   M.Kamitakahara, M.Kawashita, N.Miyata, T.Kokubo and T.Nakamura, "Bioactivity and mechanical properties of polydimethylsiloxane (PDMS)-CaO-SiO$_2$ hybrids with different PDMS contents", J.Sol-Gel Sci.Tech., 21, 75-81 (2001).

(9)   T.Kokubo, H.Kushitani, S.Sakka, T.Kitsugi and T.Yamamuro, "Solutions able to reproduce in vivo surface-structure changes in bioactive glass-ceramic A-W", J.Biomed.Mater.Res., 24, 721-734 (1990).

(10) T.Kokubo, "Bioactive glass ceramics: propertiesand applications", Biomaterials, 12, 155-163 (1991).

(11) T.Iwamoto, K.Morita and J.D.Mackenzie, "Liqiud state ^{29}Si-NMR study on the sol-gel reaction mechanisms of ormosils", J.Non-Cryst.Solids. 159, 65-72 (1993).

(12) L.Tellez, J.Rubio, F.Rubio, E.Morales, J.L.Oteo, "Synthesis of inorganic-organic hybrid materials from TEOS, TBT and PDMS", J.Mater.Sci., 38, 1773-1780 (2003).

IN VITRO COMPARISON OF THE APATITE INDUCING ABILITY OF THREE DIFFERENT SBF SOLUTIONS ON Ti6Al4V

Sahil Jalota, A. Cuneyt Tas, and Sarit B. Bhaduri
School of Materials Science and Engineering,
Clemson University, Clemson, SC 29634, USA

ABSTRACT

Coating of titanium-based biomedical devices with a layer of carbonated, apatitic calcium phosphate (CaP) increases their bone-bonding ability. Synthetic or simulated body fluids (SBF) have the ability of forming apatitic calcium phosphates on the immersed titanium alloys within few days to 2 weeks. Apatite-inducing ability of 5 M NaOH-etched surfaces of Ti6Al4V strips (10 x 10 x 1 mm) were tested by using three different SBF solutions all concentrated by a factor of 1.5. SBF solutions used in this comparative study were i) 4.2 mM HCO_3^- TRIS-HCl buffered SBF (conventional SBF or c-SBF), ii) 27 mM HCO_3^- TRIS-HCl buffered SBF (Tas-SBF), and iii) 27 mM HCO_3^- HEPES-NaOH buffered SBF (revised SBF or r-SBF). c-SBF (4.2 mM HCO_3^-) was quite slow in forming CaP on Ti6Al4V strips after 1 week of soaking at 37°C, whereas Tas-SBF of 27 mM HCO_3^- was able to fully coat the immersed samples. Cell viability, protein concentration and cell attachment were tested on the coated Ti6Al4V strips by using mouse osteoblasts (7F2).

INTRODUCTION

SBF solutions are able [1-3] to induce apatitic calcium phosphate formation on metals, ceramics or polymers (with proper surface treatments) immersed in them. SBF solutions, in close resemblance to the original Earle's (EBSS) [4] and Hanks' Balanced Salt Solution (HBSS) [5], were prepared to simulate the ion concentrations of human blood plasma. EBSS, which has 26 mM of HCO_3^- and a Ca/P molar ratio of 1.8, should be considered as a close ancestor of today's SBF solutions [3]. HBSS solution has a Ca/P ratio of 1.62. EBSS and HBSS solutions are derived from the physiological saline first developed by Ringer in 1882 [6]. It was recently reported that HBSS solutions are also able to slowly induce apatite formation on titanium [7], due to its low Ca/P ratio. For mimicking the ion concentrations of human blood plasma, SBF solutions have relatively low Ca^{2+} and HPO_4^{2-} concentrations of 2.5 mM and 1.0 mM, respectively [8]. pH values of SBF solutions were fixed at the physiologic value of 7.4 by using buffers, such as TRIS (tris-hydroxymethyl-aminomethane)-HCl [3] or HEPES (2-(4-(2-hydroxyethyl)-1-piperazinyl)ethane sulphonic acid)-NaOH [9, 10]. The buffering agent TRIS present in conventional SBF (c-SBF) formulations, for instance, was reported [11] to form soluble complexes with several cations, including Ca^{2+}, which further reduces the concentration of free Ca^{2+} ions available for the real time calcium phosphate coating. To the best of our knowledge, this behavior has not yet been reported for HEPES. HCO_3^- concentration in SBF solutions has been between 4.2 mM (equal to that of HBSS) [1] and 27 mM in revisited SBFs [12-14]. c-SBF, which was first popularized by Kokubo in 1990 [1], can be regarded as a TRIS/HCl-buffered variant of HBSS, whose Ca/P molar ratio was increased from 1.62 to 2.5. HBSS and c-SBF solutions have the same low carbonate ion concentrations (i.e., 4.2 mM). Tas et al. [12, 13] was the first in 1999 to raise the carbonate ion concentration in a TRIS-HCl buffered SBF solution to 27 mM, while Bigi et al. [9] have been the first to do the same in a HEPES-NaOH buffered SBF solution. Table I summarizes these SBF solutions. Eagle's minimum

111

Table I

Table I   Ion concentrations of human plasma and synthetic solutions (mM)

	Blood plasma	Ringer[6]	EBSS[4]	HBSS[5]	c-SBF[1]	Tas-SBF[12]	Bigi-SBF[9]	r-SBF[10]
Na^+	142.0	130	143.5	142.1	142.0	142.0	141.5	142.0
K^+	5.0	4.0	5.37	5.33	5.0	5.0	5.0	5.0
Ca^{2+}	2.5	1.4	1.8	1.26	2.5	2.5	2.5	2.5
Mg^{2+}	1.5		0.8	0.9	1.5	1.5	1.5	1.5
Cl^-	103.0	109.0	123.5	146.8	147.8	125.0	124.5	103.0
HCO_3^-	27.0		**26.2**	4.2	4.2	**27.0**	**27.0**	**27.0**
HPO_4^{2-}	1.0		1.0	0.78	1.0	1.0	1.0	1.0
SO_4^{2-}	0.5		0.8	0.41	0.5	0.5	0.5	0.5
Ca/P	2.5		1.8	1.62	2.5	2.5	2.5	2.5
Buffer					TRIS	TRIS	HEPES	HEPES
pH	7.4	6.5	7.2-7.6	6.7-6.9	7.4	7.4	7.4	7.4

essential medium (MEM) [15] and Dulbecco's phosphate buffer saline (PBS) [16], which are used in cell culture studies, may also be added to this table.

Dorozhkina et al. [17] studied the influence of HCO_3^- concentration in SBF solutions and concluded that "increasing the HCO_3^- concentration in c-SBF from 4.2 to 27 mM resulted in the formation of homogeneous and much thicker carbonated apatite layers." The same fact was also reported by Kim et al. [14] on PET substrates immersed into r-SBF. Dorozhkina et al. [17] emphasized that HEPES was rather unstable, in comparison to TRIS, and it easily lost some of the initially present dissolved carbonates. Kokubo et al. [10], who developed the HEPES-buffered r-SBF recipe, also reported that r-SBF would release CO_2 gas from the fluid, causing a decrease in HCO_3^- concentration, and an increase in pH value, when the storage period was long. Furthermore, they clearly stated that r-SBF would not be suitable for long-term use in the biomimetic coating processes owing to its instability [10]. To accelerate the SBF-coating processes, solutions equal to 1.5 times the ionic concentration of SBF were often used [2]. The aim in coating otherwise bioinert materials (such as, PET [14] or PTFE [18]) should have been the formation of bonelike, carbonated (not greater than 6 to 8% by weight) calcium phosphate layers with Ca/P molar ratios in the range of 1.55 to 1.67 [19].

The in vitro apatite-inducing ability of neither Tas-SBF [12] nor r-SBF [10] has yet been reported on Ti6Al4V substrates, in direct comparison to c-SBF. The motivation for the present study stems from our interest in finding experimental evidence to the following questions: (a) do the use of different buffers (TRIS or HEPES) in SBFs cause remarkable changes in the morphology or thickness of the calcium phosphate coat layers formed on Ti6Al4V? (b) does the variation in HCO_3^- concentration affect the apatite-inducing ability of SBFs? and (c) how would the in vitro tests with mouse osteoblast discriminate between CaP coatings of different SBFs?

EXPERIMENTAL PROCEDURE

Ti6Al4V strips (Grade 5, McMaster-Carr), with the dimensions of 10 x 10 x 1 mm, were used as substrates. The strips were first abraded with a #1000 SiC paper (FEPA P#1000, Struers), and then washed three times, respectively with acetone, ethanol, and deionized water in an ultrasonic bath. Each one of such strips was then immersed in 50 mL of a 5M NaOH solution at 60°C for 24 hours in a glass bottle, followed by washing with deionized water and drying at 40°C.

Details of SBF preparation routines are given in Table II. Freshly prepared, 1.5x c-SBF [1], Tas-SBF [12], and r-SBF [10] solutions were used in coating experiments.

		Table II	1.5 x SBF preparation	
		Weight (g per L)		
Order	Reagent	c-SBF[1]	Tas-SBF[12]	r-SBF[1, 10]
1	NaCl	12.0540	9.8184	8.1045
2	NaHCO₃	0.5280	3.4023	1.1100
3	Na₂CO₃	—	—	3.0690
4	KCl	0.3375	0.5591	0.3375
5	K₂HPO₄.3H₂O	0.3450	—	0.3450
6	Na₂HPO₄	—	0.2129	—
7	MgCl₂.6H₂O	0.4665	0.4574	0.4665
8	1 M HCl	15 mL	15 mL	—
9	HEPES	—	—	17.8920
10	CaCl₂.2H₂O	0.5822	0.5513	0.5822
11	Na₂SO₄	0.108	0.1065	0.1080
12	TRIS	9.0945	9.0855	—
13	1 M HCl	50 mL	50 mL	—
14	1 M NaOH	—	—	0.8 mL

NaOH-treated Ti6Al4V strips were soaked at 37°C in 50 mL of 1.5x c-SBF, Tas-SBF and r-SBF in tightly sealed Pyrex® bottles of 100 mL-capacity, for a period of 7, 14 and 21 days. All the SBF solutions were replenished at every 48 hours. Strips were removed from the SBF solutions at the end of respective soaking times, and washed with deionized water, followed by drying at 37°C. The strips were placed either "horizontally" on the base of the immersion bottles or dipped "vertically" into the solutions with a stainless steel wire. Coated strips were examined by using an X-ray diffractometer (XDS 2000, Scintag Corp., Sunnyvale, CA), operated at 40 kV and 30 mA with monochromated Cu K_α radiation. X-ray data were collected at 2θ values from 4° to 40° at a rate of 0.01° per minute. FTIR analyses were performed directly on the coated strips (Nicolet 550, Thermo-Nicolet, Woburn, MA). Surface morphology of the sputter-coated (w/Pt) strips was evaluated with a scanning electron microscope (FE-SEM; S-4700, Hitachi Corp., Tokyo, Japan).

Mouse osteoblast cells, designated 7F2 (ATCC, Rockville, MD), were used for cell attachment studies on the SBF-coated strips. Cells were first grown at 37°C and 5% CO_2 in alpha MEM, augmented by 10% FBS. The culture medium was changed every other day until the cells reached a confluence of 90-95%. Osteoblasts were seeded at a density of 10^5 cells/cm². Cell cytotoxicity measurements were carried out after 24 hours, cell viability assessment was performed after 72 hours and total protein amount were measured after 7 days. Adhesion of the cells was quantified 24 hours after seeding. Trypan blue was added and the cells were counted using an Olympus BX60 light microscope. Only cells that stain blue were deemed necrotic because of plasma membrane damage. For statistics, all experiments were performed in triplicate where n=3. Analysis of variance was performed using the Tukey-Kramer multiple comparisons test. Osteoblast morphology after attachment was further examined using SEM. Prior to SEM investigations, samples were soaked in the primary fixative of 3.5% glutaraldehyde. Further, the cells were dehydrated with increasing concentrations of ethanol (50%, 75%, 90% and 100%) for

113

10 minutes each. Critical drying was performed according to the previously published procedures [20]. Samples were sputter-coated with Pt prior to the SEM imaging at 5 kv.

RESULTS AND DISCUSSION

During our preliminary studies, we also prepared 1xSBF solutions (i.e., c-, r-, and Tas-SBF) and tested the formation of calcium phosphates (CaP) on alkali-treated Ti6Al4V strips for 1 week of soaking at 37°C. There was almost no coating observed, regardless of the replenishment rate with these 1xSBF solutions. For 1xSBF solutions, more than 3 weeks of soaking is required to observe only the onset of coating. To accelerate the coating process, 1.5xSBF solutions were then prepared.

Ti6Al4V strips were placed vertically (i.e., the strips placed at the halfway point along the entire height of the solution level) in the 1.5xSBF-immersion bottles. Vertical placement of the strips produced a uniform, precipitate-free coating on the surfaces, in all three SBF solutions tested. Vertically-placed strips were also coated on both sides. SEM micrographs given in Figures 1(a) through 1(c) showed the uniform CaP coatings obtained.

c-SBF solutions with 4.2 mM HCO_3^- yielded a thin layer of calcium phosphate (CaP) coating, Fig. 1(a), in comparison to Tas-SBF solutions, Fig. 1(b). The nano-morphology difference between the CaP coatings of Tas- and r-SBF solutions was quite significant, Figs. 1(b) and 1(c). Tris-solutions always produced round globules consisting of 'needle-like, intermingling, interlocking nanosize calcium phosphates, whereas Hepes-solutions produced spherical aggregates on Ti6Al4V strips.

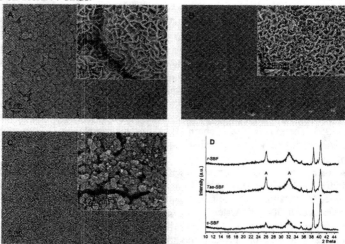

Fig. 1   Vertically-soaked Ti6Al4V strips, 1 week; (a) c-SBF, (b) Tas-SBF, (c) r-SBF, (d) XRD data; A denoted peaks of apatitic CaP, * Ti6Al4V peaks

XRD traces of the vertically-placed, 1 week-coated Ti6Al4V strips are given in Figure 1(d). c-SBF solutions still formed a lesser quantity of apatitic CaP on the strips as compared to those formed by the Tas- and r-SBF. These data clearly showed that (a) the carbonate ion concentration in 1.5xSBF solutions of pH 7.4 must be raised to the level of human blood plasma, i.e., 27 mM, to form a coating layer which fully covers the available strip surface in about 1 week, (b) the geometrical placement of the samples in SBF solutions has a strong effect on the morphology of the coatings, and (c) nano-morphology of the coatings obtained in HEPES-buffered r-SBF solutions were significantly different than those obtained in TRIS-buffered c- and Tas-SBF solutions. The initial progress of the SBF-coating, as a function of soaking time, on vertically-placed Ti6Al4V strips was also studied. SEM micrographs of Figures 2(a) to 2(d) demonstrated the morphology differences between the CaP coatings of Tas- and r-SBF solutions, both having 27 mM HCO_3^-, after 2 and 4 days of soaking. FTIR data of the 3 weeks-soaked samples showed that all the coatings were consisted of carbonated (CO_3^{2-} ion absorption bands seen at 1470-1420 and 875 cm^{-1}) calcium phosphates, Figure 2(e). The absence of the stretching and the vibrational modes of the O-H group at 3571 and 639 cm^{-1} confirmed [21] that these coatings cannot simply be named as "hydroxylapatite." From the FTIR data alone, it is rather difficult to distinguish between the coatings of different SBF solutions.

The degree of supersaturation for carbonated apatite plays a major role in determining the coating behavior of a SBF solution and is directly proportional to the activity of the individual ion as calculated out by Lu et al. [22]. Under ideal solution conditions, the activity of the individual ion is equal to its concentration in the solution. Thus increasing the concentration of the individual ion will increase the activity and thus, increase the level of supersaturation. Another parameter affecting the behavior of coating is the ionic strength. c- and Tas-SBF have the same ionic strength values i.e., 160.5 mM, whereas r-SBF has a significantly lower value of 149.5 mM. Theoretically, if a solution has a low ionic strength, this means that the ionic diffusion will be enhanced in such a solution. Thus, in a solution with low ionic strength and high ionic diffusion, there will be more nucleation sites for the precipitation reactions, which will follow. CO_2 release from an aqueous solution will also be faster in a low ionic strength solution.

Mouse osteoblasts showed significant differences in terms of the number of attached cells, cell viability, and protein concentration, as presented in Figures 2(f) through 2(h), between the apatitic calcium phosphate layers obtained by using the SBF solutions of this study. The number of attached cells, %viability, and protein concentration were all found to yield the highest values in the case of using a 27 mM HCO_3^--containing, TRIS-HCl buffered SBF solution (i.e., Tas-SBF).

Osteoblast attachment on the surfaces of the SBF-coatings (on 3 weeks-soaked samples) was monitored by SEM, and given in Figures 3(a) through 3(f). Osteoblast behavior is sensitive to biochemical and topographical features (i.e., microarchitecture) of their substrate. The ideal and most preferred surface used by osteoblasts *in vivo* is the osteoclast resorption pit [23]. However, one may speculate that the surfaces of nanoporous, apatitic CaP coatings formed in an SBF solution at 37°C and pH 7.4 represents the next-to-the-best 'bioceramic' substrate for the osteoblasts to respond to. The cytotoxicity, % viability and the protein content histograms given in Figures 2(f) to 2(h) showed that the CaP-coated (in Tas-SBF) Ti6Al4V strips always performed better than either bare Ti6Al4V or NaOH-treated TiAl4V strips. Mouse osteoblasts were able to differentiate between CaP coatings of different SBF solutions. It was quite osteoblasts to respond to. The cytotoxicity, % viability and the protein content histograms given in Figures 2(f) to 2(h) showed that the CaP-coated (in Tas-SBF) Ti6Al4V strips always performed better than either bare Ti6Al4V or NaOH-treated TiAl4V strips. Mouse osteoblasts were able to differentiate between CaP coatings of different SBF solutions. It was quite interesting to note in Figure 2(h) that the adsorbed protein concentration measured in 7-days soaked samples of Tas-SBF was even higher than those of 21-days soaked in c-SBF solution. It s a well-known fact that the surface chemistry of a material determines the initial *in vitro*

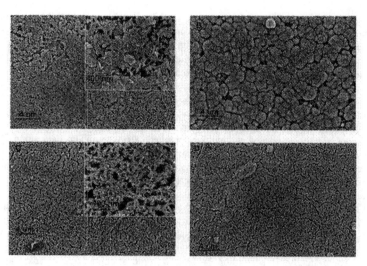

Fig. 2   (*a*) 2 days in *Tas*-SBF, (*b*) 4 days in *Tas*-SBF, (*c*) 2 days in *r*-SBF, (*d*) 4 days in *r*-SBF

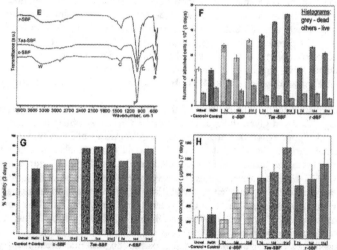

Fig. 2   (*e*) FTIR data of coatings; W: water, C: carbonate, P: phosphate bands, (*f*) number of osteoblasts attached on
the CaP coatings of different SBFs after 3 d, (*g*) cell viability for *c*-, *Tas*- and *r*-SBF coatings after 3 d, (*h*)
protein concentrations for different SBFs after 7 d.

interactions of proteins, such as fibronectin with integrin cell-binding domains, which in turn
regulate the cell adhesion process. On coating surfaces, cells were flattened and spread with clear
actin fibers associated with vinculin adhesion plaques (Fig. 3). The SEM micrograph of Fig.
3(d), recorded on a *Tas*-SBF-coated Ti6Al4V surface, clearly showed the actin cytoskeleton and
the stress fibers. Micrographs of 3(b) and 3(f) displayed the vinculin adhesion plaque formation

on samples produced by using *c*-SBF and *r*-SBF, respectively. Cells are seen to produce fewer adhesion plaques while still in the process of migration than once permanently settled in place. Sun *et al.* [24] exposed cells to calcium phosphate particles and reported that HA particles (100 nm) or β-TCP particles (100 nm) inhibited the growth of primary rat osteoblasts, while causing an increase in their expression of alkaline phosphatase. In addition, Pioletti *et al.* [25] observed a decrease in growth, viability, and synthesis of extracellular matrix in primary rat osteoblasts that were exposed to β-TCP particles (1–10 μm) or $CaHPO_4 \cdot 2H_2O$ particles (1–10 μm). The absence of such effects in our case proved the biocompatible nature of the SBF-coatings studied. Figures 3(a), 3(c) and 3(e) revealed the extension of osteoblast filopodia.

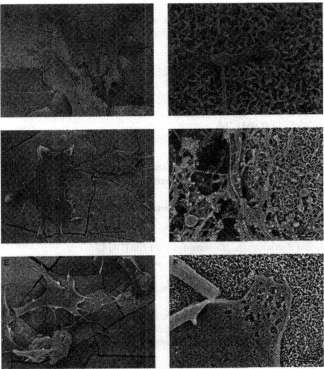

Fig. 3   SEM micrographs for the osteoblast attachment/adhesion on CaP coatings (3 weeks of soaking time) of different SBF solutions, (*a*) – (*b*): *c*-SBF, (*c*) – (*d*): *Tas*-SBF, (*e*) – (*f*): *r*-SBF.

CONCLUSIONS

This study enabled the direct comparison of HEPES and TRIS-buffered SBF solutions with one another, as well as with those of different HCO_3^- concentration and supersaturation levels. Apatite-inducing ability of NaOH-treated Ti6Al4V strips were compared when these strips were soaked at 37°C, from 2 days to 3 weeks, in three different SBF solutions, namely, *c*-SBF, *r*-SBF, and *Tas*-SBF. Although *r*-SBF and *Tas*-SBF both match the HCO_3^- concentration of human blood plasma, i.e., 27 mM, the former was buffered with HEPES, while the latter with

117

TRIS. The findings are: 1) There is a significant difference in coating morphology between the r- and Tas-SBF solutions both of 27 mM HCO_3^-. r-SBF formed more solution precipitates in comparison to TRIS-buffered SBF solutions (Tas-SBF); that would limit its long term use in coating substrates. 2) The nominal HCO_3^- concentration of an SBF solution is of crucial importance; an SBF solution containing 4.2 mM HCO_3^- can not contend in coating Ti6Al4V surfaces with an SBF solution of 27 mM HCO_3^-. The apatitic CaP formation rate of c-SBF on NaOH-treated Ti6Al4V was inferior to that of Tas-SBF. 3) SBF-coating process was considerably affected by the placement geometry of substrates in the SBF solutions; horizontally-placed substrates exhibited a growth pattern extending upwards in the solution, while the vertically-soaked strips were coated uniformly. 4) In $vitro$ tests with rat osteoblasts cultured on the apatitic CaP coatings of this study favored the TRIS-buffered, 27 mM SBF solutions in terms of osteoblast attachment, cell viability and protein concentrations.

REFERENCES

1. T. Kokubo, *J. Non-Cryst. Solids* **120** (1990) 138.
2. T. Kokubo, *Acta Mater.* **46** (1998) 2519.
3. T. Kokubo, H.-M. Kim, M. Kawashita and T. Nakamura, *J. Mater. Sci. Mater. M.* **15** (2004) 99.
4. W. Earle, *J. N. C. I.* **4** (1943) 165.
5. J. H. Hanks and R. E. Wallace, *Proc. Soc. Exp. Biol. Med.* **71** (1949) 196.
6. S. Ringer, *J. Physiol.* **4** (1880-1882) 29.
7. L. Frauchiger, M. Taborelli, B. O. Aronsson and P. Descouts, *Appl. Surf. Sci.* **143** (1999) 67.
8. H.-M. Kim, H. Takadama, F. Miyaji, T. Kokubo, S. Nishiguchi and T. Nakamura, *J. Mater. Sci. Mater. M.* **11** (2000) 555.
9. A. Bigi, E. Boanini, S. Panzavolta and N. Roveri, *Biomacromolecules* **1** (2000) 752.
10. A. Oyane, K. Onuma, A. Ito, H.-M. Kim, T. Kokubo and T. Nakamura, *J. Biomed. Mater. Res.* **64A** (2003) 339.
11. A. P. Serro and B. Saramago, *Biomaterials* **24** (2003) 4749.
12. D. Bayraktar and A. C. Tas, *J. Eur. Ceram. Soc.* **19** (1999) 2573.
13. A. C. Tas, *Biomaterials* **21** (2000) 1429.
14. H.-M. Kim, T. Kokubo, J. Tanaka and T. Nakamura, *J. Mater. Sci. Mater. M.* **11** (2000) 421.
15. H. Eagle, *Science* **122** (1955) 501.
16. R. Dulbecco and M. Vogt, *J. Exp. Med.* **106** (1957) 167.
17. E. I. Dorozhkina and S. V. Dorozhkin, *Coll. Surface. A*, **210** (2002) 41.
18. L. Grondahl, F. Cardona, K. Chiem, E. Wentrup-Byrne and T. Bostrom, *J. Mater. Sci. Mater. M.* **14** (2003) 503.
19. P. A. A. P. Marques, M. C. F. Magalhaes and R. N. Correia, *Biomaterials* **24** (2003) 1541.
20. T. J. Webster, R. W. Siegel and R. Bizios, *Biomaterials* **20** (1999) 1221.
21. C. K. Loong, C. Rey, L. T. Kuhn, C. Combes, and M. J. Glimcher, *Bone*, **26** (2000) 599.
22. Xiong Lu and Yang Leng, *Biomaterials* **26** (2005) 1097.
23. B. D. Boyan, Z. Schwartz, C. H. Lohmann, V. L. Sylvia, D. L. Cochran, D. D. D. Dean and J. E. Puzas, *J. Orthop. Res.* **4** (2003) 638.
24. J. S. Sun, Y. H. Tsuang, C. J. Liao, H. C. Liu, Y. S. Hang and F. H. Lin, *J. Biomed. Mater. Res.* **37** (1997) 324.
25. D. P. Pioletti, H. Takei, T. Lin, P. V. Landuyt, Q. J. Ma, S. Y. Kwon and K. L. P. Sung, *Biomaterials* **21** (2000) 1103.

IN SITU AND LONG TERM EVALUATION OF CALCIUM PHOSPHATE CEMENT BEHAVIOR IN ANIMAL EXPERIMENT

Masashi Mukaida[1], Masashi Neo[1], and Takashi Nakamura[1]
[1]Department of Orthopedic Surgery Graduate School of Medicine, Kyoto University, 54 Kawahara-cho, Shogoin, Sakyo-ku, Kyoto, 606-8507 Japan

Yasutoshi Mizuta[2], Yasushi Ikeda[2], and Mineo Mizuno[2]
[2]Japan fine Ceramics Center, 2-4-1, Mutsuno, Atsuta-ku, Nagoya, 456-8587 Japan

ABSTRACT

High-resolution X-ray CT is a powerful means of analyzing a comprehensive range of ceramic biomaterials in vivo. The benefit of this method is that morphological and volume changes of implant materials can be evaluated without retrieval of the implant, allowing animal viability to be maintained and allowing long-term repeated evaluation.

In this study, in situ techniques for the observation of calcium phosphate cement (CPC) were developed. CPC was implanted into the femur and under the skin of rats. The volume and morphology change of the CPC were repeatedly measured in the same rats for more than 12 months.

The structure of the CPC was visualized in three dimensions (3-D), and its volume was quantified using 3-D structure analysis software, which enabled two-value processing and estimation of the quantities of the CPC. Moreover, some CPC samples were retrieved and observed by SEM.

The surface of the CPC changed from smooth to jagged as time increased. The volume of CPC implanted into bone gradually decreased with time. The volume loss was 8% after 12 months. The volume of the CPC implanted subcutaneously increased by 7% in one month, and subsequently decreased gradually.

HRXCT was found to be a powerful means for analyzing biomaterials such as porous ceramics and bone cements in vivo.

INTRODUCTION

High-Resolution X-ray 3-D CT (HRXCT), using a microfocus X-ray tube with a high-precision computing system for 3-D reconstruction and analysis, has a spatial resolution of about 5 μm and allows the true 3-D structure of bone to be assessed by nondestructive analysis. The 3-D structures of materials are imaged and reconstructed from hundreds of 2-D sectional CT images, which are obtained at one time by a 360-degree rotation of the sample. The structures of materials are visualized in 3-D and analyzed using 3-D structure analysis software. We have established this technique for the examination of the porous structure of ceramic biomaterials [4]. This system is very useful for reconstructing and analyzing the porous structure of these biomaterials and the formation of new bone in the material. An additional benefit of this method is that morphological and volume changes of implant materials can be evaluated without retrieval of the implant from an animal body, allowing animal viability to be maintained and allowing long-term repeated evaluation for more than a year.

Time-dependent and quantitative analysis of calcium phosphate cement (CPC) is difficult, because the filling of implanted sites with injected plastic cement is difficult to estimate quantitatively. In this study, we investigated the rate and mechanism of resorption of CPC using HRXCT of living animals. CPC was implanted into the femur and under the skin of rats. The volume and morphology change of the CPC were repeatedly measured in the same rats for more than 12 months.

This is the first reported study using a High-Resolution X-ray 3-D CT system for living animals in this field.

MATERIALS AND METHODS

Calcium phosphate cement

Calcium phosphate cement was made from BIOPEX (Mitsubishi Pharma, Osaka, Japan). The cement powder consists of 75% α-tri-calcium phosphate (TCP), 5% di-calcium phosphate di-hydrate (DCPD), 18% tetra-calcium phosphate (TeCP), and 2% hydroxyapatite (HA). Following procedures reported by Kurashina et al. [1,2], the powder was mixed for 1 min with water containing sodium succinate (12%) and sodium chondroitin sulfate (5%) as described in the manufacturer's instructions. In this study, a powder:liquid ratio of 3.3 was chosen for ease of mixing and to provide sufficient strength as a filling material [3].

120

Animals and CPC graft

Sixteen mature Wistar rats, with an average weight of about 300 g, were used. The animals were husbanded, and experiments were performed, at the Institute of Laboratory Animals, Faculty of Medicine, Kyoto University.

The rats were anesthetized with sodium pentobarbital injected intraperitoneally (50 mg/kg body weight). A 15 mm skin incision was made on the anteromedial aspect of the knee, and a hole 3 mm in diameter was drilled retrogradely in the bilateral femoral intercondylar notch. The drilling was performed under saline irrigation to minimize thermal necrosis of the adjacent bone tissue. After removal of the cortical and underlying trabecular bone, the hole was irrigated with saline. The calcium phosphate cement, prepared as described above, was inserted into the hole and was also implanted subcutaneously in a prehardened form. The holes and the wounds were closed in layers. All surgery was performed under standard aseptic conditions.

High-Resolution X-ray 3-D CT (HRXCT)

The HRXCT equipment used in this work was composed of two types of microfocus X-ray tube. The focal point sizes are about 8 μm and 4 μm for ANDREX (now YXLON) MX-4 and HAMAMATSU PHOTONICS C8033, respectively. The two-dimensional image detector was an image intensifier (TOSHIBA 9 in) with a 1.3 megapixel digital CCD camera, which has a resolution of 7 lp/mm for X-ray images. The computer system controlling and reconstructing the CT was a BIRPXS-ACTIS+3, which was composed of a dual Pentium III processor (850 MHz) and VOLUME CT software.

The volume CT scanning and analysis were performed on the rats under anesthesia just after surgery and at 1, 3, 6, and 12 months after surgery (Figure 1). Figure 2 shows the 2-D CT image of the CPC implanted into the femur, taken by HRXCT. By using multi-cross-sectional CT images, a 3-D CT image could be reconstructed using software. Figure 3 illustrates an example of a 3-D image of CPC.

Histological examinations

Histological examinations were made. Undecalcified femurs were examined by scanning electron microscopy (SEM).

Figure 1. HRXCT systems for observing a living rat.

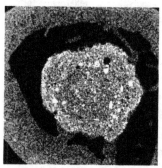

Figure 2.   2-D CT image of CPC implanted in bone.

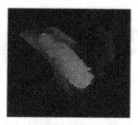

(a) 0 month   (b) 3 month   (c) 12 month

Figure 3.   3-D CT images of CPC in the bone of a rat.

RESULTS

Analysis of the 3-D CT images was performed using software, which provides volume measurements for the specimens. The surface of the calcium phosphate cement implanted into bone changed from smooth to jagged as time increased. New bone formed around, and bonded directly to, the CPC. The bonding area between the new bone and CPC increased with time. The volume of CPC implanted into bone gradually decreased with time and was 92.2% of the initial volume 12 months after implantation (Figure 4a). Figure 3 shows an example of a 3-D image of the CPC, illustrating that the volume gradually decreased with time.

The CPC implanted subcutaneously was 107.2% of its initial volume one month after implantation, and after that its volume gradually decreased (Figure 4b). At 12 months after implantation, its volume was 95.7% of its initial volume, and there was no new bone formation around the CPC. The morphology of the CPC implanted into subcutaneous tissue became partially jagged, and there were some small fragments of CPC in the peripheral area.

SEM analysis indicated that the new bone formed around the CPC implanted into bone had bonded directly to the CPC. There was partial replacement of CPC by new bone (Figure 5). The ingrowing replacement of the CPC with new bone increased with time.

DISCUSSION

Time-dependent quantitative analysis of CPC is difficult because filling of implanted sites with injected plastic cement is difficult to estimate quantitatively. To investigate the resorption of, and bone formation in, calcium phosphate cements, the cements have previously been implanted in the form of prehardened cylinders with a well-defined geometry [5]. To analyze the specimens, animals needed to be sacrificed and assessment made by destructive analysis. However, by using HRXCT for living animals, we can investigate CPC implants and their subsequent hardening in situ and can assess them without their retrieval from the animals' bodies, allowing animal viability to be maintained and allowing long-term repeated evaluation.

We have established this technique for the examination of the porous structure of ceramic biomaterials. This system is very useful for reconstructing and analyzing these porous structures. Using software, three-dimensional HRXCT images can be analyzed to quantify pore and path size distributions and the formation of new bone on the material. A benefit of this method is that morphological and volume changes of implant materials can be evaluated quantitatively by nondestructive analysis.

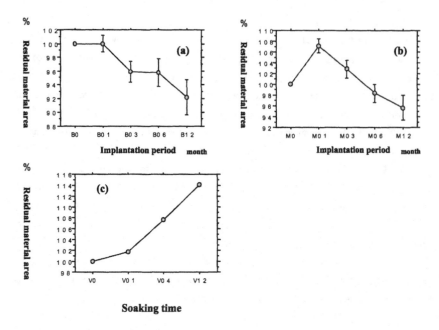

Figure 4. Relative volume of CPC vs. time of implantation or soaking:

(a) implanted into bone, (b) implanted into subcutaneous tissue, (c) soaked in SBF

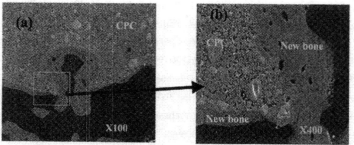

Figure 5. SEM images of CPC implanted for one month in bone of a rat :

(b): partly enlarged image (a)

124

The CPC volume in bone gradually decreased with time. The volume loss was around 8% after 12 months. SEM analysis indicated that new bone formed around, and bonded to, the CPC directly, and that there is partial replacement of CPC by new bone. There may be a remodeling process of the CPC implant into bone.

The volume of the subcutaneous CPC implant increased by 7% after one month. After that, its volume gradually decreased. XRD patterns of CPC soaked in a simulated body fluid (SBF) showed high HA peaks [3], and its surfaces were covered with apatite precipitate. This may explain the increased volume of the subcutaneously implanted CPC. Using HRXCT, the volume of CPC soaked in an SBF increased by 7% after one month. However, the volume of CPC soaked in an SBF increased with soaking time (Figure 4c). SEM analysis indicated that CPC soaked in an SBF was covered with apatite precipitate (Figure 6a). The architecture of the apatite precipitate (Figure 6b) was similar to the peripheral architecture of the CPC implanted into subcutaneous tissue at one month (Figure 7b). This precipitation of apatite may cause the increase in volume observed in CPC implanted into subcutaneous tissue.

CONCLUSIONS

We can investigate CPC implants and their subsequent hardening in situ using HRXCT, which allows their quantitative assessment in vivo without retrieval, allowing animal viability to be maintained and allowing long-term repeated evaluation.

The volume loss of CPC implanted into bone was 8% after 12 months. The volume of CPC implanted subcutaneously increased by 7% after a month and subsequently decreased gradually.

HRXCT was found to be a powerful means for analyzing biomaterials such as porous ceramics and bone cements in a living body.

ACKNOWLEDGEMENT

This work was supported in part by the National Research & Development Programs for Medical and Welfare apparatus entrusted by the New Energy and Industrial Technology Development Organization (NEDO) to the Japan Fine Ceramics Center.

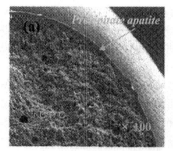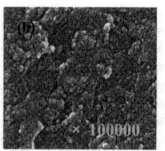

Figure 6. SEM images of CPC soaked in SBF :

(a) with the CPC covered with apatite precipitate,

(b) architecture of the apatite precipitate.

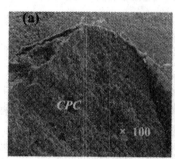

Figure 7. SEM images of CPC implanted into subcutaneous tissue:

(a) at one month,

(b) peripheral architecture of the CPC.

REFERENCES

[1] K. Kurashina, H. Kurita, A. Kotani, H. Takeuchi, M. Hirano, "In vivo study of a calcium phosphate cement consisting of α-tricalcium phosphate/dicalcium phosphate dibasic/tetracalcium phosphate monoxide", Biomaterials 18,147-51 (1997)

[2] K. Kurashina, H. Kurita, M. Hirano, de Belieck JMA, Klein CPAT, de Groot K, "In vivo study of calcium phosphate cements: implantation of an α-tricalcium phosphate/dicalcium phosphate dibasic/tetracalcium phosphate monoxide cement paste", Biomaterials 18, 539-43 (1997)

[3] K. Kurashina, H. Kurita, M. Hirano, de Belieck JMA, Klein CPAT, de Groot K, "Calcium phosphate cement:in vitro and in vivo studies of the α-tricalcium phosphate/dicalcium phosphate dibasic/tetracalcium phosphate monoxide system", J Mater Sci. Mater Med 6, 340-47 (1995)

[4] Y. Ikeda, Y. Mizuta, M. Mizuno, K.Ohsawa, M. Neo, T. Nakamura."3D CT analysis of porous structures of apatite ceramics and in-vivo bone formation", Am. Cerm. Soc., Proc.of 26th Cocoa beach Conf. on Adv. ceram. & Comp., in printing

[5] K. Ohura, M. Bohner, P. Hardouin, J. Lemaitre, G. pasquier, B. Flautre, "Resorption of, and bone formation from, new β-tricalcium phosphate-monocalcium phosphate cements: an in vivo study", J Biomed Mater Res30,193-200 (1996)

RESORPTION RATE TUNABLE BIOCERAMIC: Si&Zn-MODIFIED TRICACIUM PHOSPHATE

Xiang Wei
Iowa State University
Department of Materials Science and Engineering
322 Spedding Hall
Ames, IA, 50011

Mufit Akinc
Iowa State University
Department of Materials Science and Engineering,
2220C Hoover Hall
Ames, IA, 50011

ABSTRACT

An ideal bone implant material would support the activity of osteoblasts in the development of new bone, while simultaneously being resorbed by osteoclasts as part of the lifelong orderly process of bone remodeling. Silicon and Zinc modified tricalcium phosphate, a biphasic material, was synthesized as a candidate for resorbable temporal bone implant having a controlled solubility and pharmaceutical effect to promote bone formation. From XRD and ICP analyses, it was shown that up to 10mol% Si and Zn can be incorporated in tricalcium phosphate (TCP) without formation of a secondary phase. Changes in lattice parameters and unit volume of TCP as calculated by Rietveld analysis indicate that Si and Zn substitute for P and Ca respectively. The dissolution study was carried out in simulated body fluid. The chemical analysis and XRD results imply that the Si and Zn additives not only decrease the solubility of TCP, but also slow hydroxyapatite (HAp) precipitation, indicating that dissolution of temporary implant and formation of new bone may be tailored by the level of Si and Zn substitution.

INTRODUCTION

Among the calcium phosphate ceramics, tricalcium phosphate has been investigated most extensively as the primary resorbable bioceramics for bone replacements[1-3]. According to the CaO/P_2O_5 phase diagram, tricalcium phosphate (TCP) exists in three crystalline forms: the β-tricalcium phosphate (β-$Ca_3(PO_4)_2$, β-TCP) is stable below 1125°C, at which temperature it transforms to α-tricalcium phosphate (α-$Ca_3(PO_4)_2$, α-TCP), and α' (α'-TCP) is stable above 1430°C[4]. Dissolution rate of β-TCP was reported to be 3-12 times faster than stoichiometric hydroxyapatite ($Ca_5(OH)(PO_4)_3$, HAp)[5]. In vitro studies revealed that the α-TCP had a higher dissolution rate than β-TCP[3]. Ducheyne et al. compared the dissolution rates of six calcium phosphates in calcium and phosphate free solution at pH 7.3. The dissolution rate increased from HAp to tetra calcium phosphate (TTCP) in the following order[6]:

$$HAp < CDAp < \beta\text{-TCP} < \alpha\text{-TCP} < TTCP$$

Due to its higher solubility, TCP as an implant, is expected to degrade in the host and be gradually replaced by the regenerating bone. Based on different tissue and implant conditions, the biodegradability of TCP ceramics are vary widely[7,8]. TCP was reported to be more bioresorbable than HAp that usually show minimal resorption. TCP behaves as a seed of bone and a supplier of the Ca and PO_4 ions. Furthermore, α-TCP can be handled as a paste and set in situ. However, higher solubility of TCP as bone implant results in loss of strength. Thus, the suitability of TCP for use in vivo is critically dependent on how the dissolution rate might be controlled by chemical modification. The biphasic α and β-TCP with controlled dissolution behavior will be a promising resorbable biomaterial for temporary implant.

It has been shown that the solubility of zinc or magnesium doped β-TCP decreases with Zn and Mg content[9,10]. The decrease in solubility was attributed to the increased stability of the β-TCP structure caused by the addition of Zn or Mg ions. Langstaff et al found that Si-doped HAp formed modified HAp and α-TCP phases following sintering[11,12]. This material was stabilized in biological media and could be resorbed when acted upon by osteoclasts[11-13]. In their studies, β-TCP, HAp and commercial HAp showed ~0.02% calcium dissolution per day, whereas α-TCP and Si-HAp was ~0.03% and ~0.006% respectively. No report was found on the effect of these additives on α and β-TCP biphasic materials.

From a pharmaceutical point of view, Zn and Si are osteoconductive and could stimulate osteogenesis for bone growth. Zinc polycarboxylate cements have also been used in dentistry for many years[14]. Since zinc oxide (ZnO) was able to form salt-bridges between Zn and carboxylate ions, a simply blended mixture of ZnO/HAp was used to form the bone cement with polyacrylic acid aqueous solution. The effect of a silicon deficient diet on chicks indicated that silicon is an essential trace element for the normal growth and development of chicks[15]. An increase in dietary silicon has been directly linked to an increase in bone mineralization. Electron microprobe analysis and imaging ion microscopy showed that silicon is localized in sites of active bone formation in young rats and mice[16].

MATERIALS AND METHOD

Si,Zn-modified TCP and pure α-TCP was prepared by sintering $CaCO_3$, $NH_4H_2PO_4$, ZnO and fumed SiO_2 at certain mole ratio at 1300°C and quenching in dry air. The β-TCP was produced by sintering at 1000°C. More preparation details were described in another paper[17].

Chemical composition of the powders was determined by the ICP (Thermo Elemental, Franklin, MA). An X-ray powder diffractometer (Scintag Inc. CA) with Cu Kα radiation was used for crystal structure analysis and lattice parameter determinations. The Rietveld analysis of the diffraction patterns with different additive levels was performed by Rietica software. Initial crystal parameters for α and β-TCP were taken from the literature[18,19].

Four different samples were investigated for dissolution behavior in synthetic body fluid: synthesized α-TCP, β-TCP, Si,Zn-TCP-5 and Si,Zn-TCP-10 (the number 10 refers to 10 mol% of P and Ca are replaced by Si and Zn respectively). The glass bottles were sealed and kept in a water bath shaker in order to preserve the temperature at 37°C. The pH of the solution was measured with an accuracy of ±0.02. Suspensions were centrifuged and Ca concentration of clear solution was measured by atomic absorption spectroscopy (Perkin-Elmer 5000). The standard solutions for AAS were prepared based on the same ion concentration of SBF except Ca^{2+}. The solid was washed several times by de-ionized water and dried in vacuum for XRD and SEM.

RESULTS
Composition and Structure

Table 1 lists the nominal and measured calcium, phosphorus, zinc and silicon contents for synthesized α-TCP and Si,Zn-TCP powders as determined by ICP. The accuracy of analysis was 3% with a reproducibility of 5%. The measured values are very close to the intended compositions indicating negligible composition change during synthesis. The ratio of (Zn+Ca):(Si+P) is round 1.5, same as the ration of stoichiometric TCP.

Table 1 Chemical analysis of the synthesized TCP samples

Samples	Si [wt%]		Zn [wt%]		Ca:P*	(Zn+Ca):(Si+P)*
	Nominal	Measured	Expected	Measured	Measured	Measured
α-TCP	0	0.008	0	0.099	1.53	N/A
Si,Zn-2+	0.36	0.35	1.25	1.27	1.487	1.457
Si,Zn-10	1.77	1.97	6.15	6.03	1.477	1.433

+Si,Zn-2: TCP modified by the addition of 2 mol% each of Si and Zn.
*Expressed as mole ratio

X-ray diffraction patterns of sintered Si,Zn-TCP show excellent agreement with the published JCPDS files for α-TCP (#29-395), β-TCP(#09-169), or mixture of the two depending on additives content (Fig. 1). No evidence for other crystalline phases, such as $CaSiO_3$ or other phosphates, was observed indicating that Zn and Si are dissolved completely in the TCP structure. Figure 1 also compares XRD patterns of Si,Zn-5, Si-5, and Zn-5 samples. Si,Zn-5 shows a mixture of α and β phases, while Zn-5 has only the β phase and Si-5 has α as the major phase implying that Zn addition favors formation of β phase, and will increase the phase transformation temperature ($T_{\beta \to \alpha}$); while Si addition favors α structure, which will decrease $T_{\beta \to \alpha}$. The phase distribution analysis calculated by the Rietveld refinement method show that α:β ratio decreases from about 9:1 to 2:3 as the concentration of additives increased from 1 to 5 mol%, indicating that Zn addition dominates the phase composition. In the phase diagrams of $Ca_3(PO_4)_2$-Ca_2SiO_4, the solubility limit of Si in TCP is approximately at a molar ratio of 4 mol% expressed as $Si/(Si + P)$ [20]. In the Si-5 sample, silicocarnotite appears as the second phase.

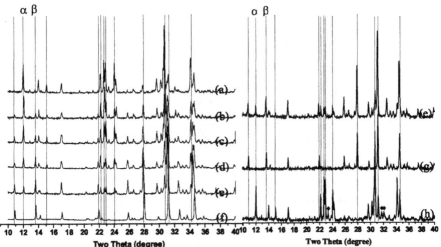

Fig. 1 XRD patterns of synthesized Si,Zn-TCP with different doping level. (a)Si,Zn-1 (b)Si,Zn-2 (c) Si,Zn-3 (d)Si,Zn-4 (e)Si,Zn-5 (f)Si,Zn-10 (g)Zn-5 (h)Si-5. The dot lines represent the distinct peaks of α-TCP, and the dash lines are β-TCP. ◆ is silicocarnotite phase.

Si,Zn-TCP samples show that up to 10 mol% Si and Zn can be incorporated into the TCP lattice without formation of a second phase.

XRD patterns of Si,Zn-TCP show slight shift, indicating a change in the lattice parameters imposed by the additives. Based on ionic radii, the substitution of Zn for Ca will lead to a contraction, while the substitution of Si for P will cause expansion of the unit cell[21]. The lattice parameters and unit cell volumes (V_{uc}) are shown in Figure 2. The effect of Zn addition on the structure is demonstrated as parameters a, c and V_{uc} of Zn-TCP were all smaller than those of pure TCP. $V_{uc}(\alpha)$ of Zn-5 is 10 Å^3 smaller than α-TCP (4307.5 Å^3 vs 4317.6 Å^3). $V_{uc}(\beta)$ of Zn-5 and Zn-10 are 45 and 314 Å^3 smaller than that of β-TCP. $V_{uc}(\alpha)$ of Si-5 is 34Å^3 larger than that of α-TCP as expected from Si substitution. Compared to α-TCP, the lattice parameters of Si-TCP and Si,Zn-TCP expand both in b and c directions implying that Si substitutes primarily in sites promoting expansion along these directions. Si,Zn-TCP samples showed competitive effect of Si and Zn: $V_{uc}(\alpha)$ of Si,Zn-TCP samples are larger than that of the pure α-TCP, but decrease at higher doping level. $V_{uc}(\beta)$ of Si,Zn-TCP is smaller than pure β-TCP but larger than Zn-TCP. Hence zinc ion has the dominant effect on the β phase, and silicon has a dominant effect on the α phase. At higher doping levels, Zn dominates the structural changes brought by the additives.

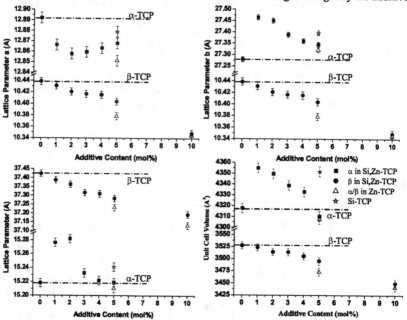

Fig. 2 Lattice parameters and unit cell volume of Si,Zn-TCP determined by Lebail refinement. The error bars are based on standard deviation of triplicate measurements.

Dissolution Behavior

Variation of [Ca^{2+}] as a function of time in SBF in contact with TCP powders is given in Figure 3. For α-TCP, [Ca^{2+}] shows an increase in the first couple of days then decreased at longer time with slightly increasing of pH. This behavior may be explained by an initial dissolution of Ca^{2+} followed by reprecipitated in the form of HAp (HAp is the least soluble of the phosphates). The β-TCP shows similar dissolution behavior as α-TCP, but the lower Ca^{2+} concentration indicates the lower solubility than α-TCP. In contrast, Ca^{2+} concentration of SBF containing Si,Zn-TCP-10 didn't change much with constant pH, indicating that the additives inhibited not only the dissolution of TCP, but also the precipitation of HAp. For Si,Zn-TCP-5, Ca^{2+} concentration increases slightly for the first several days then decreases, $[Ca^{2+}]$ is between TCP and Si,Zn-TCP-10. The decrease in solubility can be attributed to the increased stability of the TCP structure as a result of the addition of Zn and Si. XRD patterns of the samples after aging in SBF show that the solid phase of α-TCP sample was almost entirely converted to HAp, while no HAp phase was observed in Si,Zn-TCP samples (Fig. 4).

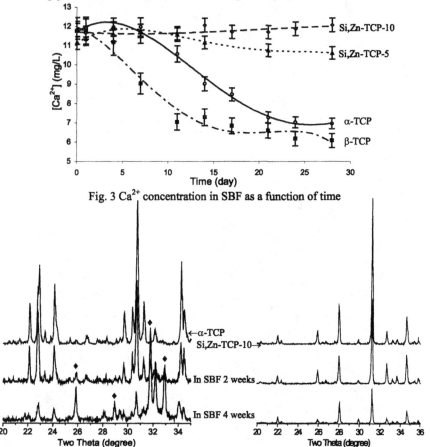

Fig. 3 Ca^{2+} concentration in SBF as a function of time

Fig. 4 XRD patterns of α-TCP and Si,Zn-TCP in SBF at different soaking time. ◆ is HAp phase.

133

After immersion in SBF, the surface morphology of Si,Zn-TCP was significantly different from the α-TCP (Fig. 5). The surface of α-TCP after 4 weeks exposure to SBF shows characteristic solution-precipitated needle-like morphology of HAp. TEM micrograph shows thin fiber like features, radiating from the center out. The specific surface area of α-TCP increased from 0.47 to 51.47 m^2/g in 4 weeks due to the needle-like crystal morphology and smaller particle size. The XRD patterns and SEM micrographs of β-TCP also indicated the formation of small amount of HAp after soaking in SBF. However, there is no obvious morphology change in Si,Zn-TCP samples, which also prove that Si and Zn inhibited not only the dissolution but also the precipitation of HAp.

Fig. 5 Micrographs of α-TCP and Si,Zn-TCP in SBF. (a)synthesized α-TCP, (b)α-TCP in SBF after 4 weeks, (c) TEM micrograph of α-TCP in SBF 4 weeks (d)synthesized Si,Zn-TCP-10, (e) Si-Zn-TCP-10 in SBF after 4 weeks.

DISCUSSION

From the crystallographic point of view, the basic crystal structure of TCP can supply a suitable environment for Zn and Si substitution. In the β-TCP unit cell, Ca(4) and Ca(5) sites are unique compared to other three sites. Ca(4) is on the 3-fold axis and has an unusual coordination to the O(9), O(9'), O(9'') face of the P(1)O$_4$ group. Electrostatic repulsion between the cation and the phosphorus atom is expected to be significant and the Ca(4)$\cdots$O(9) bonds are longer (3.041(1) Å) than normal Ca$\cdots$O bond, which is about 2.4Å in agreement with the Pauling's rule[19,22]. The Ca(5) site have six-fold octahedral coordination surrounded by oxygens, and all six Ca$\cdots$O distances are relatively short, falling into the range 2.238-2.287Å. Therefore, these two sites are very suitable for the smaller cation, but are highly constrained for a Ca^{2+} ion. Substitution of smaller Zn^{2+} for Ca^{2+} results in more stable β-TCP structure than the undoped one by reducing the strain in the structure. In whitlockite, Mg^{2+} ions with ionic radius 0.57Å substituted Ca(4) and Ca(5) positions. The Mg(4)$\cdots$O(9) bonds were 2.907Å, shorter than that of Ca(4)$\cdots$O(9) bonds. The approach of O$\cdots$Mg(5)$\cdots$O angles toward 90° with increasing Mg content provides further confirmation of the trend toward a more ideal octahedral configuration[22]. Similarly, tetrahedral PO_4^{3-} ions in the TCP structure may be replaced by SiO_4^{4-} units, resulting in Si substituted TCP. According to the structure, each formula unit occupies 180 Å^3 in α-TCP compared with 168 Å^3 in the β form[23]. Since α-TCP has a more open structure than β-TCP, the larger Si^{4+} ions favor α over β structure.

The charge compensation due to Si^{4+} substitution for P^{5+} may be explained either by oxygen vacancies, or additional proton (H$^+$) incorporation into the structure. In contrast to expanded lattice parameter b and c, the contracted lattice parameter a of Si,Zn-TCP may result from the oxygen vacancies in this direction. A more detailed crystal structural analysis of Si-Zn-TCP is necessary to elucidate the mechanism of charge compensation.

CONCLUSION

Si,Zn-modified TCP exhibits α-, β-TCP structure, or mixture of the two depending on the level of the additives. The changes in lattice parameters and unit cell volume clearly demonstrate that Si and Zn are structurally incorporated into TCP and stabilize the structure. Zn can substitute for the Ca and causes a contraction of the unit cell. Si can substitute for the P and results in expansion of the unit cell. 10 mol% addition of Si & Zn appears to prevent dissolution of TCP, and inhibit precipitation of HAp. By varying the Zn & Si additions, the dissolution behavior of TCP may be controlled.

REFERENCES

[1] RZ LeGeros, JP LEGeros, G Daculsi, R Kijkowaka, Encyclopedic Handbook of Biomaterials and Bioengineering. Calcium phsophate biomaterials: preparation, properties, and biodegradation, ed. E.R. Schwartz. 1995, New York: Marcel Decker. 1429-63.

[2] C.P.A.T. Klein, A.A. Driessen, and K.d. Groot, Relationship between the degradation behavior of calcium phosphate ceramics and their physical-chemical characteristics and ultrastructural geometry. Biomaterials, 1984. 5(3): p. 157-160.

[3] J.B. Park and R.S. Lakes, Biomaterials: An Introduction 2nd ed. 1992, New York: Plenum Publishing.

[4] ER. Kreidler and FA. Hummel, Phase relationships in the system SrO-P$_2$O$_5$ and the influence of water vapor on the formation of Sr$_4$P$_2$O$_9$. Inorg. Chem, 1967. 6: p. 884-891.

[5]M. Jarcho, Calcium phosphate ceramics as hard tissue prosthetics. Clin Orthop Rel Res, 1981. **157**: p. 259-78.

[6]P. Ducheyne and S. Radin. Bioceramics, ed. W. Bonfield and G.W. Hastings. Vol. 4. 1991, London: Butterworth-Heinemann. 135-144.

[7]C.P.A.T. Klein, et al., Biodegradation behavior of various calcium phosphate materials in bone tissue. J. Biomed Mater Res, 1983. 17: p. 769-84.

[8]F. Driessens, Formation and stability of calcium phosphate in relation to the phase composition of the mineral in calcified tissues. Bioceramic of calcium phosphates, ed. D.G. K. 1983, Boca Raton: CRC Press. 1-32.

[9]I. Manjubala, preparation of biphasic calcium phosphate doped with magnesium fluoride for osteoporotic applications. J. Mater. Sci. letter, 2001. 20: p. 1225-1227.

[10]A. Ito, Resorbability and solubility of zinc-containing tricalcium phosphate. J. Biomed Mater Res, 2002. 60: p. 224-231.

[11]S. Langstaff and M. Sayer, Resorbable bioceramics based on stabilized calcium phosphates. Part I: rational design, sample preparation and materials characterization. Biomaterials, 1999. 20: p. 1727-1741.

[12]S. Langstaff and M. Sayer, Resorbable bioceramics based on stabilized calcium phosphates. Part II: evaluation of biological response. Biomaterials, 2001. 22: p. 135-150.

[13]I.R. Gibson, S.M. Best, and W. Bonfield, Chemical characterization of silicon-substituted hydroxyapatite. J. Biomed Mater Res, 1999. 44: p. 422-428.

[14]D. Xie, et al., A hybrid zinc-calcium-silicate polyalkenoate bone cement. Biomaterials, 2003. 24: p. 2794-2757.

[15]E.M. Carlisle, Silicon: a requirement in bone formation independent of vitamin D1. Calcif Tissue Int., 1981. 33(1): p. 27-34.

[16]W.J. Landis, D.D. Lee, and J.T. Brenna., Detection and localization of silicon and associated elements in vertebrate bone tissue by imaging ion microscopy. Calcif Tissue Int., 1986. 38(1): p. 52-9.

[17]X. Wei and M. Akinc, Si,Zn-modified tricalcium phosphates: A. phase composition and crystal structure study. Key Engineering Materials, 2005. 284-286: p. 83-86.

[18]M Mathew, et al., The crystal structure of α-Ca3(PO4)2. Acta Cryst., 1977. p.1325-33.

[19]B. Dickens, L.W. Schroeder, and W.E. Brown, Crystallographic Studies of the Role of Mg as a Stabilizing Impurity in β-tricalcium phosphate: I. The Crystal Structure of Pure β-tricalcium phosphate. J. Solid state Chemistry, 1974. 10: p. 232-248.

[20]R.W. Nurse, J.H. Welch, and W. Gutt, High-temperature phase equilibrium in the system of dicalcium silicate-tricalcium phosphate. J. the Chem. Soc., 1959: p. 1077-1083.

[21]D.R. Lide, CRC Handbook of Chemistry and Physics 84th edition. 2003, Boca Roton: CRC Press LLC.

[22]L.W. Schroeder, B. Dickensand, and W.E. Brown., Crystallographic Studies of the Role of Mg as a Stabilizing Impurity in b-tricalcium phosphate: II. Refinement of Mg-containing β-tricalcium phosphate. J. Solid state Chemistry, 1977. 22: p. 253-262.

[23]J.C. Elliott, Structure and Chemistry of the Apatites and other Calcium Orthophosphates. 1994, London: Elsevier.

Dental Ceramics

MICROLEAKAGE OF A DENTAL RESTORATIVE MATERIAL BASED ON BIOMINERALS

Håkan Engqvist, Emil Abrahamsson, Jesper Lööf and Leif Hermansson
Doxa AB
Axel Johanssons gata 4-6
SE-751 51 Uppsala
Sweden

ABSTRACT

Since the introduction of the resin composites on the market marginal leakage causing secondary caries has been one of the major clinical topics. As the composites shrink during hardening they tend to develop a gap between the filling and the tooth, where bacteria can enter. This can be overcome by using bonding techniques, but bonding does not give complete success. An alternative to use shrinking composites could be the use of fully ceramic filling material based on biominerals that harden via a acid base reaction with water. The biomineral technology based on Ca-aluminate has been proven to yield a bond to living tissue and as such the filling material would naturally avoid marginal leakage without the use of any pre-treatment or extra bonding systems. In this paper the influence of thermo cycling on the marginal leakage of a restorative material based on biominerals is evaluated and compared to that of a resin composite.

INTRODUCTION

Marginal leakage and related possible caries is the most common reason for replacement of a filling. More than 50 % of all dental restorations today are restorations of old dental fillings [1]. This can be correlated to the contact zone between the filling material and tooth structure, possible bonding materials and the tooth surface structure and stresses in the contact zone. A new chemically bonded ceramic system has been proposed as an alternative material to those based on polymers or metals [2]. Special interest has been shown the $CaO-Al_2O_3$ – system [3-5]. Since the calcium aluminate technology provides a material with both a small expansion and bioactivity it is hypothesized that it would result in minimal marginal leakage [6-7]. The micro-leakage is normally tested *in vitro* via thermo-cycling of fillings in extracted teeth and evaluation of the contact zone between the filling and tooth structure by dye penetration techniques [8]. The aim of this investigation is to study the marginal leakage of Class V experimental Ca-aluminate material (Doxa) fillings and compare the results with those of a resin composite.

MATERIALS AND METHODS

Materials and materials preparation

A Ca-aluminate based material (Doxa Experimental) was used in this study and as a reference a commercial resin-based composite material system (filling and bonding) was employed and used according to instructions. The chemical oxide composition of the experimental material expressed as oxides is shown in Table I. The main active phase is the

1:1 phase of Ca-aluminate system, CaOxAl$_2$O$_3$, named Marokite. This was synthesised by heat treatment of the basic oxides at 1400 ^{0}C.

Table I. Chemical composition of the Doxa experimental material

Oxide parts	Wt-%
Ca-aluminate, Marokite	46
Inert glass	31
Water	19
Others	< 5

The microstructure of the experimental Doxa material is shown in figure 1. The small white spots are glass particles, the light-grey phase katoite (Ca$_3$[Al(OH)$_4$]$_2$(OH)$_4$ and the dark-grey phase gibbsite (Al(OH)$_3$).

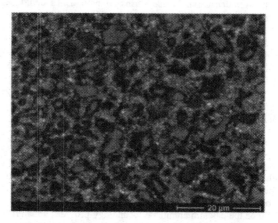

Figure 1. SEM micrograph of the experimental Doxa material.

A typical property profile of the material is shown in Table II.

Table II. Some property data for the Doxa experimental material

Hardness (Hv 100 g)	120
Compressive strength (MPa)	240
Flexural strength (MPa)	82
Initial setting (min)	3
Final setting (min)	6
Shear bond strength (MPa)	10

Eight extracted teeth were divided into two groups. In group I Doxa experimental fillings were prepared and in group II the resin composite fillings. Four cavities were prepared in each tooth. Standardised Class V preparations (5 mm wide and 3 mm deep) were made at the cemento-enamel junction on available buccal, lingual, mesial and distal surfaces. All preparations were etched with Conditioner 36 (Dentsply) for 15 seconds. All fillings in one tooth were done at the same time. The tooth was placed in 100% RH for 10 minutes followed by storage in a phosphate buffer system (PBS) for 7 days.

Experimental procedure

The restorations were ground with a straight cylinder (Horico Diamond medium). To check the materials tendency to show leakage the fillings were subjected to 500 cycles in 5 and 55 °C water with 30 s "hold time". The dye penetrant (Spotcheck; Magnaflux Corp) was then applied to the restoration surface and remained on the surface for 30 seconds before excess dye was removed. Dye penetration images of the unfilled, filled and stressed cavities were taken using a stereo-microscope (LOM) with a digital camera connected to a frame grabber program. The microstructural evaluation of the bulk material and the contact zone was done using scanning electron microscopy (SEM).

RESULTS AND DISCUSSION

The unstressed Doxa experimental filling in light microscopy is shown in figure 2.

Fig. 2. Image of an unstressed Doxa experimental filling (LOM).

None of the Doxa fillings showed any dye penetration (n=16) after the thermo-cycling program. The marginal leakage could be considered as zero, see the figures below.

141

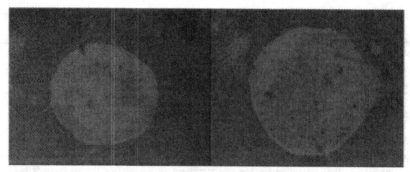

Figs. 3- 4. Images of stressed Doxa experimental fillings (LOM). Note that no marginal leakage could be detected but the surface structure is changed compared to the unstressed fillings (Fig. 2).

However, as can be seen in the figures 3-4 the surface structure of the fillings changed somewhat. As has been earlier reported the system in pure water will exhibit a surface reaction or precipitation upon the surface [9]. The exact composition of the surface was not determined but it is reasonable to believe, based on the earlier findings, that it is composed of calcite originating from the water bath used during stressing. No dye penetration could be detected on the cross-sectioned fillings. For the composite, 15 out of 16 fillings showed micro-leakage, see Figs. 4-5. Most of the marginal leakage occurred towards the enamel side of the cavity.

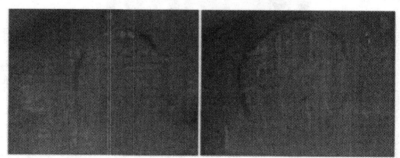

Figs. 4-5. Images of stressed resin composite fillings (LOM).

The complete filling of the gaps between the experimental Doxa filling and tooth structure resulting in zero micro-leakage is basically related to the curing mechanism of the material, which involves dissolution of the Ca-aluminate during reaction with water, and a precipitation of nano-size hydrates on the tooth structure. This mechanism of dissolution-precipitation is repeated during the hardening of the material. A magnification of the contact region between the Ca-aluminate based material and enamel is shown in figure 6. The gap between the restorative material and the enamel is filled with hydrates. The white particles are glass particles.

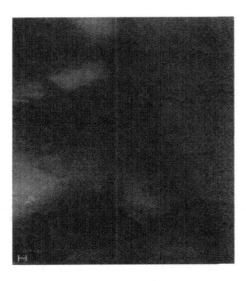

Fig 6. SEM micrograph of the contact zone between the Doxa restorative
material and the enamel (to the right) – magnification 40.000 X.

Gaps between particles within the material as well as voids towards the tooth structure is
filled with the hydrates formed in the over all reaction

$$3\ CaO \cdot Al_2O_3\ +\ 12\ H_2O\ \rightarrow\ Ca_3[Al(OH)_4]_2(OH)_4\ +\ 4Al(OH)_3$$

Ca-aluminate   water   Katoite   Gibbsite

A slight expansion of < 0.2 linear-% contributes also to the sealing of the contact zone. The
reaction mechanisms have been described in more details in [1,6-7].

CONCLUSIONS

The experimental Doxa filling did not show any marginal leakage. The complete sealing
between the material and the tooth is due to the hardening mechanism. Some spots on the
surface of the filling showed due penetration. However this could not be seen on cross-
sectioned filling surfaces. In this study the reference material – a commercial resin based
material showed extensive micro-leakage.

143

REFERENCES

[1] I. Mjör et al, International Dental Journal (2000) Vol 50 No 6, p

[2] L Kraft, Ph D Thesis: "Calcium Aluminate Based Cement as Dental Restorative Materials" Faculty of Science and Technology, Uppsala University, Sweden. 2002

[3] L. Hermansson, L. Kraft, H. Engqvist. "Chemically Bonded Ceramics as Biomaterials". Proceedings of the 2nd International Symposium on Advanced Ceramics, Key Engineering Materials vol 247 (2003) pp. 437-442. 2nd ISAC, 2002, Shanghai, China.

[4] J. Loof, H. Engqvist, L. Hermansson and N-O Ahnfelt, Mechanical Testing of Chemically Bonded Bioactive Ceramic Materials, Key Eng. Materials Vols. 254-256 (2004) ,51-54N.

[5] J. Loof, H. Engqvist, G. Gomez-Ortega, H. Spengler, N-O Ahnfelt and L. Hermansson, Mechanical Property Aspects of a Biomineral Based Dental Restorative System, Key Eng. Materials Vols 284-286 (2005) 741-44

[6] L. Kraft, L Hermansson, Hardnesss and Dimensional Stability of a Bioceramic Dental Filling based on Calcium Aluminate Cement, 26th Annual Conference on Composites and Advanced Ceramics, Cocoa Beach Florida, Am. Ceram Soc, Vol 23B Issue 2002 Ed by H-T Lin and M Singh

[7] H. Engqvist, J-E. Schultz-Walz, J. Loof, G. A. Botton, D. Maye , M. W. Pfaneuf, N-O.Ahnfelt , L. Hermansson "Chemical and Biological Integration of a Mouldable Bioactive Ceramic Material capable of forming Apatite in vivo in Teeth". Biomaterials vol 25 (2004) pp. 2781-2787

[8] E Sung, S. M. Chan, E. T. Tai and A. Caputo, Effects of various irrigation solutions on microleakage of Classs V composite resorations, J Prosthetic Dentistry, Vol 91, No 3, 265-267 (March 2004)

[9] N. Axén, L. M. Bjursten, H. Engqvist, N-O Ahnfelt, L. Hermansson , Zone formation at the interface between Ca-aluminate cement and bone tissue environment Presented at 9th Ceramics: cells and tissue, Faenza, Italy Oct 2004, to be published

144

A COMPARATIVE STUDY OF THE MICROSTRUCTURE –PROPERTY RELATIONSHIP IN HUMAN ADULT AND BABY TEETH

I.M. Low, N. Duraman, J. Fulton, N. Tezuka
Materials Research Group, Dept. of Applied Physics, Curtin University of Technology, Perth, WA 6845, Australia

I.J. Davies
Materials Research Group, Dept. of Mechanical Engineering, Curtin University of Technology, Perth, WA 6845, Australia

ABSTRACT
    The structure-property relationship in human adult and baby teeth was characterised by grazing-incidence synchrotron radiation diffraction, optical and atomic-force microscopy, in addition to Vickers indentation. Similarities and differences between both types of teeth have been highlighted and discussed.   The depth profiling of hardness indicated a gradual change in microhardness from the enamel to dentine, thus confirming the graded nature of human teeth. Vickers hardness of the enamel was load-dependent but load-independent in the dentine. The use of a "bonded-interface" technique revealed the nature and evolution of deformation-microfracture damage around and beneath Vickers contacts.

INTRODUCTION
    Biological materials exhibit many levels of hierarchical structures from macroscopic to microscopic length scales, with the smallest building blocks in biological materials being generally designed at the nano-scale with nanometer-sized hard inclusions embedded in a soft protein matrix.[1] Teeth, like other natural biomaterials, are essentially inorganic/organic composites with enviable strength and damage resistant properties. In human teeth, the enamel comprises of ~96% calcium apatite, either as hydroxyapatite (HAP) ($Ca_{10}(PO_4)_6(OH)_2$) or fluorapatite ($Ca_{10}(PO_4)_6F_2$).[2-4] Such a high mineral content ensures that teeth are the hardest and probably the strongest biological material within the human body. In both the adult and deciduous tooth, enamel is the outer structure that envelops the crown. It is almost fully mineralized with highly organized HAP crystallites, making it mechanically hard and highly resistant to wear. In general, the deciduous teeth are whiter, softer, smaller, and weaker compared to their permanent counterparts. In addition, their enamel is thinner and has a higher organic content. The microstructure of enamel is highly textured with aligned prisms or rods that run approximately perpendicular from the dentin-enamel-junction (DEJ) towards the tooth surface.[2,5] Each rod consists of tightly packed carbonated hydroxyapatite crystals that are covered by a nanometre-thin layer of enamelin and oriented along the rod axis. However, in deciduous teeth, the outer-most layer is generally devoid of the usual prism structure. It remains unknown whether the HAP crystals and enamel rods are similar in dimension and distribution for both types of teeth. In contrast, dentin is the supporting structure that lies underneath enamel and is primarily composed of ~68% HAP mineralized collagenous matrix surrounding tubular extensions of the dentinoblast cells. This less mineralized tissue provides the tooth with the toughness required to resist catastrophic fracture when subjected to masticatory stresses. The DEJ is the interface region bridging across the enamel and dentine and possesses the desirable

capability of arresting crack propagation.[6-8] Hitherto, the influence of age on the structure-property relationships within both types of teeth has been poorly understood.

This paper investigates and compares the variations in crystal structure, composition, microhardness and damage within the human adult and deciduous teeth. The similarities and differences in the microstructure-property relationships of both types of teeth are highlighted and discussed.

EXPERIMENTAL METHODS

Specimen Preparation

Thin slices of adult and baby human teeth were used for the study. A precision diamond blade cutter was used to cut each tooth into two 1.0 mm thick slices that were either parallel ("occlusal-section") or perpendicular ("axial-section") to the occlusal surface. Prior to the diffraction measurement, the buccal and lingual sides were ground with SiC paper in order to obtain a plano parallel flat plate.

Both atomic-force microscopy (AFM) and optical microscopy (OM) were used to reveal the surface topography and microstructure of polished and etched tooth samples. A Nikon optical microscope and a Digital Instrument Dimension 3000 scanning probe microscope were used for the imaging study. The AFM experiments were carried out in contact mode in air with a gold-coated cantilever and a tip of Si_3N_4, with a spring constant of 0.58 Nm^{-1}.

Synchrotron Radiation Diffraction

Depth-profiling of the near-surface structure of enamel was conducted using grazing-incidence synchrotron radiation diffraction (SRD). Imaging plates were used to record the diffraction patterns at a wavelength of 0.7 Å and grazing angles, α, of 0.2, 0.4, 0.8, 1.0, 3.0, and 5.0°.

Indentation Testings

Both adult and baby tooth samples were used for the measurement of indentation responses and damage. The teeth were cut either parallel or perpendicular to the occlusal surface using a precision diamond blade. The cut specimens were then cold mounted in epoxy resin and polished to a 1 μm surface finish. Indentation responses of polished samples as a function of: (a) load (2-100 N), (b) loading time (0-24 h), and (c) depth-profile were measured using a Zwick microhardness tester. Test (a) was designed to evaluate whether or not the hardness was load-dependent due to an indentation-size effect. The viscoelastic flow or creep and graded characteristics were evaluated in test (b) and (c), respectively. The diagonal lengths of the indent, $2a$, were used to calculate the hardness, determined here as $H_v = P/2a^2$, where P is the load. Values of fracture toughness, K_{1c}, were calculated as $K_{1c} = 0.025P/c^{1.5}$ where c is the average crack length in mm.

Information of subsurface contact damage during Vickers indentations was obtained using a bonded-interface specimen configuration.[9,10] This test allows the nature and degree of damage accumulation beneath the indenter to be revealed. Polished surfaces of two half-specimens were glued face-to-face with a thin layer of adhesive under moderate clamping pressure. The top surface perpendicular to the bonded interface was then polished for the indentation tests. The two halves of the indented specimens were then separated and examined using a reflection optical microscope in Nomarski illumination.

146

RESULTS AND DISCUSSION
Composition, Microstructure and Mechanical Properties

Analysis of the structure-property relationships in both baby and adult teeth revealed several distinct similarities and differences. Firstly, the preferred grain orientation or texture in both age groups was similar with the HAP grains being aligned approximately orthogonal to each other between the occlusal- and axial-sections. These highly textured microstructures have been indicated by the SRD patterns (Figs. 1 & 2) and verified by optical and electron micrographs with elongated enamel rods being seen in the axial-section whereas key-hole shaped enamel rods were noted in the occlusal-section.[2,11]

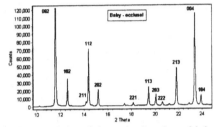

Fig. 1: SRD plot of the occlusal surface of baby tooth.

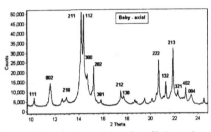

Fig. 2: SRD plot of the axial-section of baby tooth.

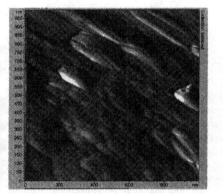

Fig. 3: AFM showing the HAP grains of an adult enamel.

Fig. 4: AFM showing the HAP grains of a baby enamel.

Secondly, phase analysis of the SRD patterns indicated HAP to be the predominant phase present in both types of teeth as indexed according to the Powder Diffraction File (PDF) 74-0566.[12] Thirdly, SRD depth-profiles of canine teeth in both age groups showed no apparent depth variations in crystallinity, composition or texture, thus indicating the presence of uniform nano- and microstructures within the tooth enamel.[11] Fourthly, the crystal size of HAP in enamel was larger in the adult specimen than in the baby specimen. This surprising size difference was

147

clearly observed by atomic-force microscopy (Figs. 3 & 4) with the mean grain size for the adult and baby enamel being 94 and 185 nm, respectively. In addition, elongated HAP grains were observed in the adult enamel whereas more equiaxed grains were found in the baby enamel. As would be expected, the size and thickness of the baby enamel was smaller when compared to that of the adult enamel.[11]

In addition, the Vickers microhardness for the baby tooth was lower than for the adult, indicating the adult tooth enamel to be harder and presumably in order to provide a greater capacity for stress-bearing and wear resistance (Fig. 5). In all cases, the hardness decreased progressively from the enamel to the dentine by virtue of a decreasing HAP content. Such a graded nature of teeth has also been previously observed.[5,13,14] Finally, when compared to the adult tooth, the baby tooth possessed a lower fracture toughness and would thus be more vulnerable to fracture (Fig. 6).

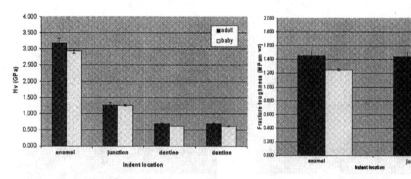

Fig. 5: Comparison of hardness profiles in the baby and adult tooth.

Fig. 6: Comparison of fracture toughness in the baby and adult tooth.

Indentation Responses and Damage

In both the baby and adult teeth, a pronounced load-dependent hardness behaviour was evident within the enamel layer (Fig. 7). In contrast, this phenomenon was not evident in the dentin layer. To the best of the authors' knowledge, the display of load-dependent hardness behaviour in human enamel is unique and has not previously been reported. This phenomenon is well-known in coarse-grained metals and ceramics such as Ti_3SiC_2 and can be attributed to a grain-size effect.[15,16] At small loads, the contact diagonal, $2a$, of the Vickers impression is smaller than the grain size, and the hardness measures properties of single grains; when $2a$ becomes much larger than the grain size at high loads, the hardness measures polycrystalline properties, with more grains oriented for deformation by slip. This indentation-size effect can be ruled out for human enamel by virtue of its fine sub-micrometre mirostructure. Instead, the origin of load-dependent hardness in human enamel may be attributed to its highly textured microstructure which favours the stochastic nature of deformation damage by virtue of a statistical variation in crystallographic orientation of individual grains. Only those grains of correct orientation will favour the occurrence of intergrain deformation along certain specified grain-boundaries. The expansion of the deformation zone is a result of the activation of additional grain-boundaries as the pressure intensifies within the Vickers compression-shear zone.

The non-viscoelastic nature of both dentin and enamel was indicated by the absence of reduction in hardness with prolonged indentation time (Fig. 8). In contrast to most viscoelastic polymers which exhibit indentation creep as a result of relaxation processes and viscoelastic flow,[17] both enamel and dentine were found to be creep resistant, which might otherwise lead to undesirable permanent deformation because of large contact stresses during mastication.

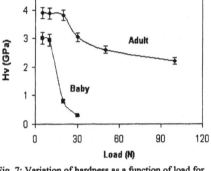

Fig. 7: Variation of hardness as a function of load for baby and adult enamel.

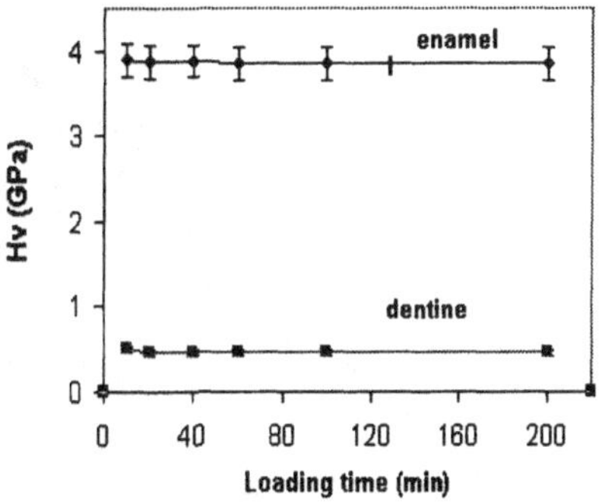

Fig. 8: Variation of hardness as a function of loading time for adult enamel and dentine.

Extensive damage was observed within the vicinity of a distorted indent at small loads in adult enamel, albeit with no indication of cracks (Fig. 9). Indentation cracks were seen to form in enamel for loads greater than 50 N but not in dentine. In the former case, extensive damage in the vicinity of the indent was also observed on the axial surface but not on the occlusal surface. In both cases, a pronounced display of anisotropy in the cracking pattern was noted which suggests a highly heterogeneous or textured microstructure. It is also interesting to note that the sides of the indent appeared to be covered with a thin layer of enamel, a characteristic exclusive to biomaterials. In contrast to this, cracks readily formed in the baby enamel even at low loads (Fig. 10) which indicates the poor fracture resistance of baby teeth.

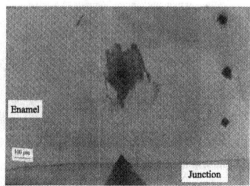

Fig. 9: Characteristics of damage in the vicinity of indents for the adult tooth.

149

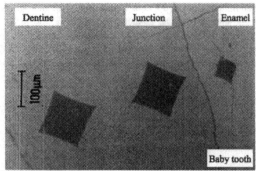
Fig. 10: Characteristics of damage in the vicinity of indents for the baby tooth.

Contact damages within the baby and adult teeth showed characteristics of quasi-plastic materials; namely, large scale compression-shear deformation and the evolution of microcracks to macrocracks as the load increased (Figs. 11 & 12). The deformation was accommodated by intragrain slip and intergrain sliding which leads to nucleation of subcritical voids or microcracks. The key to the damage tolerance lies in irreversible deformation and quasi-plasticity under conditions of intense compression-shear stresses beneath the indenter. The fully developed damage zone comprised of an accumulation of microstructurally discrete events, each consisting of some kind of intragrain shear faulting and delamination, which lead to free surface relief and intergrain microcracking. These damage processes were found to be more severe under the sharp Vickers indenter resulting in more pronounced slip, delamination, grain push-out and microcracking. Closer inspection of Fig. 12 indicated that microcracks initiate from the lamella shear faults and extend over to several grain facets. The degree of microcrack extension depends somewhat on the sign and intensity of residual thermal expansion anisotropy stresses at the grain boundary; those facets with tensile stresses will favour crack extension and vice versa for compressive facets.

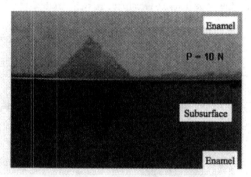

Fig. 11: Optical micrographs showing half-surface (top) and section (bottom) views of Vickers damage of an adult tooth at a low load (10 N). Note the absence of cracks on the top surface but presence of microdamage below the occlusal surface.

150

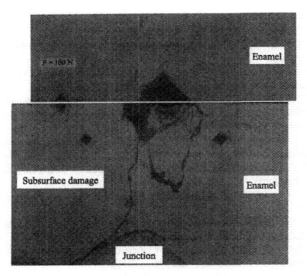

Fig. 12: Optical micrographs showing half-surface (top) and section (bottom) views of Vickers damage of an adult tooth at high load (100 N). Note the arrest of an advancing crack at the dentine-enamel junction.

CONCLUSIONS

Human adult and baby teeth exhibited distinct similarities that included: (a) HAP being the dominant phase with the enamel rods running approximately perpendicular from the dentin-enamel-junction (DEJ) towards the tooth surface; (b) Progressive decrease in hardness from enamel to dentine; (c) Hardness being load-dependent for enamel but load-independent for dentine; (d) Hardness being time-independent for both enamel and dentine; and (e) Cracks forming in enamel but not in dentine. However, when compared to the adult tooth, the baby enamel was smaller, softer, more prone to fracture, and possessed larger HAP grains.

ACKNOWLEDGMENTS

This work was performed at the Australian National Beamline Facility (ANBF) with support from the Australian Synchrotron Research Program (03/04-AB-06 & 03/04-AB-24), which is funded by the Commonwealth of Australia under the Major National Research Facilities Program. We thank Dr. J. Hester of ANBF and Dr. M. Reyhani for experimental assistance in the collection of SRD and AFM data, respectively.

REFERENCES

1. B. Ji and H. Gao, "Mechanical properties of nanostructure of biological materials," *J. Mech. Phys. Solids* **52**, 1963-1990 (2004).
2. A.R. Ten Cate, Oral *Histology: Development, Structure, and Function* (4th Edn.), 1994. Mosby, St. Louis, MO.
3. P.W. Lucas, in: B. Kurten (Ed.) *Basic Principles of Tooth Design, Teeth, Form, Function, Evolution*, 1979. Columbia Univ. Press, New York, pp. 154–162.

4. R.A. Young, "Implications of Atomic Substitutions and Other Structural Details in Apatites," J. Dent. Res. Suppl. 53, 193–203 (1974).
5. I.M. Low, "Depth-Profiling of crystal structure, texture and microhardness in a functionally-graded tooth enamel," J. Am. Ceram. Soc. 87, 2125-31 (2004).
6. Lin, C.P. and W.H. Douglas, "Structure–Property Relations and Crack Resistance at the Bovine Dentin–Enamel Junction," J. Dent. Res., 73, 1072-78 (1994).
7. Xu, H.H.K., D.T. Smith, S. Jahanmir, E. Romberg, J.R. Kelly, V.P. Thompson and E.D. Rekow. 1998. "Indentation Damage and Mechanical Properties of Human Enamel and Dentin," J. Dent. Res., 77, 472-80 (1998).
8. Low, I.M., J. Fulton, P. Cheang and K.A. Khor, "Designing New Dental Materials Through Mimicking Human Teeth," pp. 365-73 in K.A. Khor, T.S. Srivatsan, M. Wang, W. Zhou, F. Boey (Eds). Processing and Fabrication of Advanced Materials VIII. 2000. World Scientific, Singapore.
9. I.M. Low, "Vickers contact damage in micro-layered Ti$_3$SiC$_2$" J. Europ. Ceram. Soc., 18, 709 (1998).
10. I.M. Low, "A modified bonded-interface technique with improved features for studying indentation damage of materials" J. Aust. Ceram. Soc., 34, 120 (1998).
11. N. Duraman, "Mapping the structure and properties in human teeth," Physics Project 593, Curtin University of Technology, Perth, WA. (2004).
12. K. Sudarsanan, R. A. Young, "Significant precision in crystal structural details: Holly Springs hydroxyapatite," Acta Crystallogr. B, 25, 1534-1543 (1969).
13. N. Meredith, M. Sherriff, D.J. Setchell and S.A.V. Swanson. "Measurement of the Microhardness and Young's Modulus of Human Enamel and Dentin using an Indentation Technique," Arch. Oral Biology, 41, 539-545 (1996).
14. J.L. Cuy, A.B. Mann, K.J. Livi, M.F. Teaford and T.P. Weihs. "Nanoindentation Mapping of the Mechanical Properties of Human Molar Tooth Enamel," Arch. Oral Biology, 47, 281-291 (2002).
15. I.M. Low, S.K. Lee, M.W. Barsoum, B.R. Lawn, "Contact damage accumulation in Ti$_3$SiC$_2$," J. Am. Ceram. Soc. 81, 225 (1998).
16. I.M. Low, "Vickers contact damage in micro-layered Ti$_3$SiC$_2$," J. Europ. Ceram. Soc. 18, 225 (1998).
17. I.M. Low, "Effects of load and time on the hardness of a viscoelastic material," Mater. Res. Bull. 33, 1753 (1998).
18. I.M. Low, G. Paglia, C. Shi, "Indentation responses of viscoelastic materials," J. Appl. Polym. Sci. 70, 2349 (1998).

Author Index